ADVANCES IN BIOCHEMICAL ENGINEERING

Volume 9

Editors: T. K. Ghose, A. Fiechter, N. Blakebrough

Managing Editor: A. Fiechter

With 69 Figures

Springer-Verlag
Berlin Heidelberg New York 1978

ISBN 3-540-08606-4 Springer-Verlag Berlin Heidelberg New York
ISBN 0-387-08606-4 Springer-Verlag New York Heidelberg Berlin

Library of Congress Catalog Card Number 72-152360
Printed in Germany

Typesetting, printing, and bookbinding: Brühlsche Universitätsdruckerei Lahn-Gießen.
2152/3140-543210

Contents

Theory and Practice of Continuous Cultivation of Microorganisms in Industrial Alcoholic Processes

V. L. Yarovenko
All-Union Research Institute of Fermentation Products,
Moscow, USSR

Contents

1. Theoretical Principles of Continuous Culture

The continuous method of propagation of microorganisms is an important scientific achievement of technical microbiology since it can be conducted in almost constant cultivation conditions. The technical development and theoretical basis of this method can be found in the works of Monod [13, 14], Novick and Szilard [17], Herbert [6, 7], and Iyerusalimsky [8–10].

In industrial fermentation the continuous cultivation method is a most effective means of governing microbial metabolism, because it makes it possible to eliminate the lag-phase, to shorten the total time of fermentation, to use equipment more effectively, and to control process conditions automatically. The main difficulty that must be eliminated when employing continuous technology in ethanol, acetone, and butanol fermentations is contamination of the medium in which the cultivation of useful microorganisms occurs. In order to exclude foreign microflora, it is necessary to characterize the transfer processes in the battery and the physiological state of the continuous culture obtained in the following medium [20].

1.1 Bulk-Transport Processes in Multistage Batteries

For the above case, a battery is considered as consisting of N equal vessels in series, each of working volume ν [2]. The battery is initially filled with liquid A. Liquid B is continuously put into the first vessel of the battery.

The mixture of liquids flows out of the first vessel into the second one, then into the third, and so on, the liquid volume in each vessel remaining constant. It is supposed that solutions A and B have the same or approximately the same specific gravity and that in each of the vessels complete mixing of the solutions takes place. The total solids concentration in solution B is designated as S_0. While solution B is steadily flowing into the first vessel, the second, third, and subsequent ones each receive a mixture from the previous vessel in the battery.

The solids concentration in each vessel can be found on the basis of the relationship V/v, where V equals the volume of liquid entering the first vessel, assuming that at the beginning of the process the solids concentration in very vessel equals zero.

The solids concentration in the first, second, and other vessels at any time t is designated $S_1, S_2, \ldots, Sn$. After a short period of time the first vessel is charged with volume ΔV of the new liquid, i.e., $S_0 \Delta V$ weight units of solids. During this time, liquid with volume ΔV passes from the first vessel into the second. As a result, the first vessel loses $(S_1 + \theta_1 \Delta S_1)\Delta V$ weight units of solids. Here ΔS_1 is the change in solids concentration in the first vessel during period Δt, and $(S_1 + \theta_1 \Delta S_1)$ is the mean solids concentration in the mixture flowing from the first vessel into the second $(0 < \theta_1 < 1)$ during the time interval Δt. Thus, after the indicated interval, the quantity of solids in the first vessel increases by $[S_0 - (S_1 + \theta_1 \Delta S_1)]\Delta V$ weight units. Also, the increase in the quantity of solids in the first vessel during time Δt is equal to the product of liquid volume v and concentration increment ΔS_1.

Now we can deduce the material balance equation:

$$v\Delta S_1 = [S_0 - (S_1 + \theta_1 \Delta S_1)]\Delta V$$

or

$$\frac{\Delta S_1}{\Delta V} = \frac{S_0}{v} - \frac{S_1 + \theta_1 \Delta S_1}{v}.$$

In the limit as $\Delta t \to 0$, and bearing in mind that as $\Delta t \to 0$, $\Delta V \to 0$, and $\Delta S_1 \to 0$, we can deduce the following differential equation:

$$\frac{dS_1}{dV} + \frac{S_1}{V} = \frac{S_0}{V}. \tag{1}$$

Since at the beginning of the process (when V = 0) $S_1 = 0$, concentration S_1 at any moment of the process can be found by solving Eq. (1) using the initial condition (V = 0, $S_1 = 0$).
The liquid flowing from the first vessel into the second is characterized by the changeable concentration S_1, and S_1 is the above-mentioned particular solution of Eq. (1).
By analogy, a system of differential equations can be developed,

$$\frac{dS_i}{dV} + \frac{S_i}{v} = \frac{S_i - 1}{v} \quad (i = 1, 2, ..., n) \tag{2}$$

the solution of which satisfies the initial conditions and determines the solids concentration in each of the n-vessels of the battery.
Equation (2) can be solved by Laplace transformation to give:

$$S_i = S_0[1 - (1 + \frac{1}{1!}\frac{V}{v} + \frac{1}{2!}(\frac{V}{v})^2 + ... + \frac{1}{(i-1)}(V/v)^{i-1})e^{-V/v}] \tag{3}$$
$$(i = 1, 2, ..., n)$$

Equation (3) represents the solids concentration in the i-th vessel (i = 1, 2, ..., n) after filling the first vessel with volume V of solution B.
The proportion of solution B in the i-th vessel is denoted as P_i. The relation between the volume of solution B in this vessel and the vessel volume is equal to the relation between the weight of solids in the vessel and the weight of solids that would be in the vessel completely filled with solution B. Hence,

$$P_i = \frac{S_i v}{S_0 V} = 1 - [1 + \frac{1}{1!} V/v + \frac{1}{2!}(\frac{V}{v})^2 + ... + \frac{1}{(i-1)!}(V/v)^{i-1}]\, e^{-V/v} \tag{4}$$
$$(i = 1, 2, ..., n)$$

Equation (4) can be used to find the proportion of solution B in any vessel at any time. The proportion of original liquid in the i-*th* vessel at any moment can be determined by the equation:

$$R_i = 1 - P_i = [1 + \frac{1}{1!} V/v + \frac{1}{2!}(V/v)^2 + ... + \frac{1}{(i-1)!}(V/v)^{i-1}]\, e^{-V/v} \tag{5}$$
$$(i = 1, 2, ..., n).$$

The vessel cycle is defined as the period during which the volume of original liquid is replaced by the same volume of the new one. The proportion of new and old liquids in the i-th vessel after r vessel cycles are designated as $P_i^{(r)}$ and $R_i^{(r)}$, respectively.

$$P_i(r) = 1 - \left(1 + \frac{r}{1!} + \frac{r^2}{2!} + ... + \frac{r^{i-1}}{(i-1)!}\right) e^{-r} \tag{6}$$

$$R_i(r) = \left(1 + \frac{r}{1!} + \frac{r^2}{2!} + \ldots + \frac{r^{i-1}}{(i-1)!}\right) e^{-r} \quad (i = 1, 2, \ldots, n). \tag{7}$$

For a 5-vessel battery in particular the percentage of fresh liquid after r-vessel cycles is found in the following way:

$$P^{(r)} = 20\,[5 - (5 + 4r + \frac{3}{2} r^2 + \frac{1}{3} r^3 + \frac{1}{24} r^4)]\, e^{-r}. \tag{8}$$

At the same time, volume of the single vessel equivalent to volume of the 5-vessel battery is determined according to the following equation:

$$P^{(r)} = 100\,(1 - e^{r/5}). \tag{9}$$

On the basis of Eqs. (8) and (9), $P^{(r)}$ has been calculated for the two cases indicated in Fig. 1. Comparison of the results obtained shows the advantages of the battery of communicating vessels; the theoretical predictions are in good agreement with the experimental findings. After adding new liquid equal to the battery volume (r = 5), about 100% – 82,45% = 17.55% of the old liquid is retained in the battery. In the single vessel (with volume 5 v), the old liquid retention amounts to 100% – 63.21% = 36.79%. After two battery cycles the old liquid quantity amounts to 0.42% of the battery volume. In the single vessel, this value is equal to 100% – 86.47% = 13.55%.

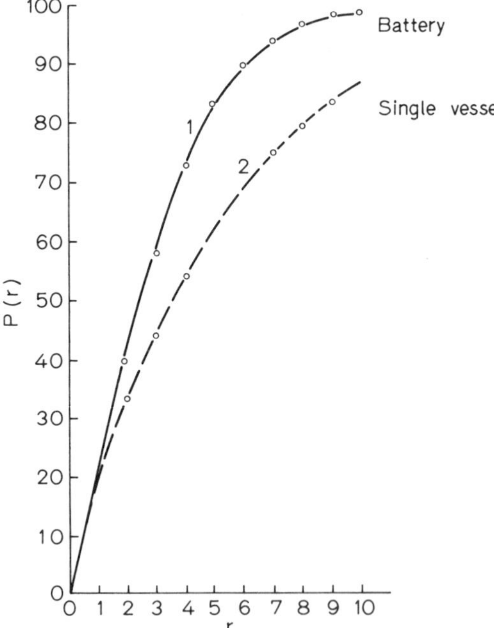

Fig. 1. The relationship between fresh liquid input and battery vessel cycles

1.2 Influence of Liquid Retention Time on the Acidity of the Medium

The liquid retention time is of great importance in continuous cultivation [20]. It can be demonstrated [2] that the mean age of the element in vessel i in a battery of i-vessels at moment τ can be determined by the equation:

$$B_i(\tau) = T \sum_{j=1}^{i} P_j^{(\tau)} \quad (i = 1, 2, ..., n) \tag{10}$$

T is vessel cycle duration and $P_j(\tau)$ is fraction of j-vessel volume (j = 1, 2, ..., i) consisting of new liquid at moment τ.
As $\lim_{\tau \to \infty} P_j(\tau) = 1$ then for large values of τ, when the process actually becomes steady-state, the asymptotic formula can be used:

$$B_i = iT \quad (i = 1, 2, ..., n). \tag{11}$$

This means that the mean age in each battery vessel under steady-state conditions remains virtually constant and equals the product of vessel cycle time and vessel cardinal number. The last n-th vessel accounts for the highest value:

$$B_n = n \cdot T. \tag{12}$$

It has been shown experimentally [21, 22], that a functional connection between mean age and beer acidity exists. Considering the more intense increase of beer acidity in the tail battery vessels and comparing the curves (Figs. 2 + 3), we have come to the conclu-

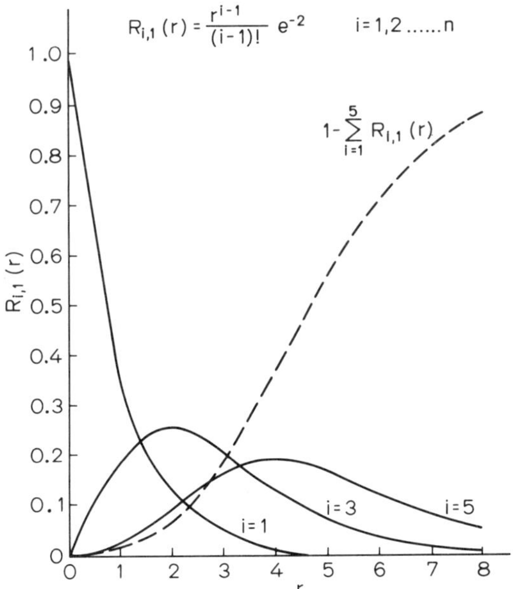

Fig. 2. The dependence of $R_i(r)$ on number of cycles

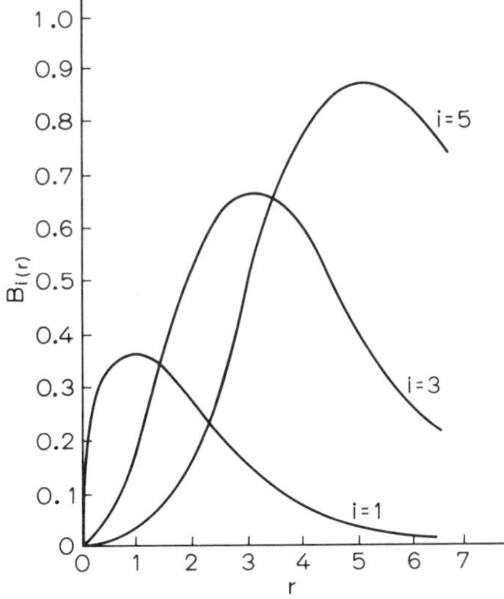

Fig. 3. The relationship between number of cycles r and the liquid retention time $B_{i(r)}$

sion that when evaluating battery capacity with respect to the degree of beer souring, it is necessary to take the retention factor into account. The higher the indicated retention factor, the higher the intensity of souring. For example, after adding samples of ripe beer to sterile wort, a notable increase in ripe beer acidity with increase in sample retention time can be observed. Figure 2 characterizes the dependence of $R_i(r)$, on the number of cycles r; Fig. 3 presents the relationship between cycle number r and the retention time $B_{i(r)}$ of the liquid which occupies the i-th vessel after r-cycles.
Hence, in order to limit acidification, it is sufficient to reduce the retention time that accounts for the contamination. From the analytical point of view, it follows that the degree of beer acidification in a continuous multistage battery is closely connected with the value of the function [3].

$$I = \sum_{i=1}^{n} \int_0^\infty R_i(r)dr \quad (i = 1, 2, ..., n). \tag{13}$$

where $R_i(r)$ is the fraction of i-th vessel volume that is occupied by the original liquid after r-cycles.
Reduction of the function I is indicative of decrease in retention time and acceleration of uptake of carbohydrates.
Next, $\int_0^\infty R_i(r)dr$ is denoted as I_i, i = 1, 2, ..., n.
Then:

$$I = \sum_{i=1}^{n} I_i. \tag{14}$$

From operational calculations it follows that:

$$\int_0^\infty R_i(r)dr = \lim_{p\to 0} LR_i \tag{15}$$

where LR_i is Laplace of function $R_i(r)$ and P is a pattern parameter.
After dividing both parts of each Eq. (2) by S_0, multiplying by v, and taking into account (4), we reach the following results:

$$\frac{dP_i}{dr} + P_i = P_{i-1} \quad i = 1, 2, ..., n \tag{16}$$

when r = 0, $P_i = 0$, $P_0 = 1$. By introducing a Lapalce unilateral transformation into every system equation and denoting the Laplace function pattern P_i as LP_i we obtain:

$$pLP_i + LP_i = LP_{i-1} \quad i = 1, 2, ..., n. \tag{17}$$

Hence:

$$LP = \frac{LP_{i-1}}{1 + p}. \tag{18}$$

Because $LP_0 = \frac{1}{p}$ (19)

then

$$LP_i = \frac{1}{p\,(1+p)^i} \qquad i = 1, 2, \ldots, n. \tag{20}$$

The fraction of the i-th volume occupied by the original liquid after r-cycles is determined by

$$R_i = 1 - P_i. \tag{21}$$

Thus:

$$LR_i = \frac{1}{p}\;1 - \frac{1}{(1+p)^i} = \frac{(1+p)^{i-1}}{p\,(1+p)^i} \qquad i = 1, 2, \ldots, n. \tag{22}$$

in particular:

$$LR_1 = \frac{1}{1+p}; \qquad\qquad LR_2 = \frac{p+2}{(1+p)^2} \tag{23}$$

Thus:

$$I_1 = \lim_{p\to 0} LR_1 = \lim_{p\to 0} \frac{1}{1+p} = 1 \tag{24}$$

$$I_2 = \lim_{p\to 0} LR_2 = \lim_{p\to 0} \frac{p+2}{(1+p)^2} = 2 \tag{25}$$

$$I = I_1 + I_2 = 3.$$

1.3 Effect of Yeast Cell Concentration on Increase of Medium Acidity

The yeast cell concentration in a battery of vessels plays a decisive role in the production of alcohol from saccharified wort.

Bringing a large quantity of seed culture into the head vessel while charging it and maintaining the yeast cell concentration at a high level hampers the increase of beer acidity and considerably accelerates the fermentation of carbohydrates. Thus, the yeast cell concentration does not decrease in the vessel when charging it.

After a certain period of time, battery vessels appear to contain equilibrium concentrations of yeast cells that are hardly influenced by the amount of seed culture introduced into the head vessel.

The yeast cell concentration in medium flowing in an i-battery vessel can be determined by the equation [20]:

$$X_i^{(p)} = \frac{1}{2}\left(X^1 - \frac{1}{kT} + \sqrt{\left(X^1 - \frac{1}{kT}\right)^2 + 4\,\frac{X_i^{(p)} - 1}{kT}}\right) \quad i = 1, 2, \ldots, n \tag{26}$$

where $X_i^{(p)} = \lim_{t \to \infty} X_i$ equilibrium concentration of yeast in the i-th fermenter,

$X_{i-1}^{(p)}$ = equilibrium concentration of yeast in beer feeding i-th fermenter,

X^1 = maximum concentration of yeast cell in original medium,

k = yeast propagation rate, and

T = duration of vessel cycle.

An important characteristic of continuous fermentation is the capacity of the battery for the degradation of sugars. The sugar-degrading capacity, designated as φ, is defined as the relationship between the substrate concentration in saccharified wort and its concentration in ripe beer [20].

Therefore, the main characteristics for comparing various means of feeding batteries include the values of function (13) for the first and second vessels (I_1 and I_2), yeast cell equilibrium concentrations in the first two vessels $X_1^{(p)}$ and $X_2^{(p)}$, and the resultant capacity φ. Their calculation in respect to different ways of battery charging [20, 3] testifies to the value of feeding in parallel to the head vessels in each battery together with recirculation of yeast separated from beer as the most efficient method. Means of recirculation are also being intensively investigated [26].

Function (13) can also be used for evaluation of various battery configurations [4].

It would be interesting to evolve an analytical description of the acidification of ripe beer in relation to liquid retention in battery vessels. The fulfillment of this task presents some difficulty, but some results could be achieved if the primary period of batch process were taken into consideration [5].

We should now closely examine the batch alcoholic process during which wort mixed with yeast is left for fermentation. Industrial fermentation rarely takes place in aseptic conditions, since some acid-producing bacteria are brought in with the malt. As a result, alcoholic fermentation proceeds concurrently with the propagation of acid-producing bacteria and beer acidification. After some time, alcohol formation practically ceases and the beer becomes ripe. But the acidity of ripe beer is not constant, and it continues to sour during storage.

The propagation of acid-producing bacteria is proportional to their concentration, but at the same time, these bacteria are known to form acids which inhibit their activity proportionally.

Let us assume that the vessel is filled at time t = 0 and that the beer acidity in °D (acid quantity) at time t is designated as P. We assume that at time $t = t_0$ alcohol formation practically ceases, and beer acidity as $P = P_0$. The number of acid-producing bacteria is denoted as X_1 at moment t. Then, in accordance with the above hypothesis, the propagation of bacteria and ripe beer acidification (Fig. 4) can be described by the following system of differential equations:

$$\begin{cases} \dfrac{dX_1}{dt} = KX_1 - K_1X_1P \\ \\ \dfrac{dP}{dt} = K_2X_1 \end{cases} \qquad (27)$$

where K, K_1, and K_2 are positive constants.

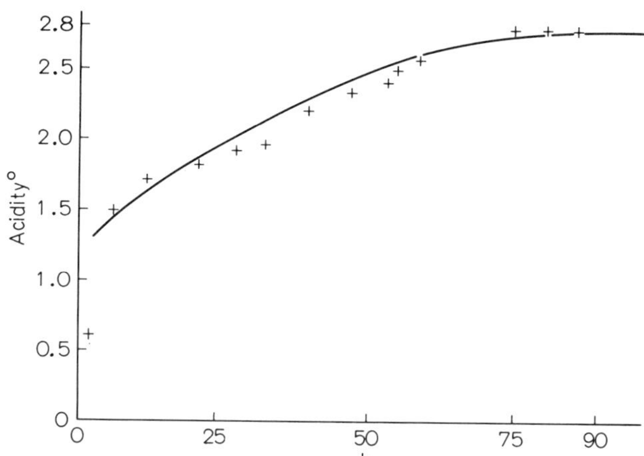

Fig. 4. The calculated acidity curve of ripe beer depending on fermentation time (the process lasts 24 h)

Now we divide the first equation by the second and obtain:

$$\frac{dX_1}{dP} = \frac{K}{K_2} - \frac{K_1}{K_2} P. \tag{28}$$

Hence:

$$X_1 = \frac{K}{K_2} P - \frac{K_1}{2 K_2} p^2 + C_1. \tag{29}$$

Since $X_1 = 0$ when $P = 0$, the integration constant $C_1 = 0$.
Thus, the relationship between bacterial number and beer acidity is expressed by the following equation:

$$X_1 = \frac{K}{K_2} P - \frac{K_1}{2 K_2} p^2. \tag{30}$$

Combining Eq. (30) with Eq. (27) a new differential equation is obtained:

$$\frac{dP}{dt} = KP - \frac{K_1}{2} p^2 \tag{31}$$

which under the indicated conditions ($t = t_0$, $P = P_0$) determines ripe beer acidity as a function of the process duration (t).
By separating the variables and integrating the equation we obtain:

$$P = \frac{K}{\frac{K_1}{2} + C_2 e^{-Kt}} - \frac{K_1}{2} + C_2 \exp_0^{-Kt}. \tag{32}$$

The integration constant C_2 is found by using the initial data:

$$C_2 = \left(\frac{K}{P_0} - \frac{K_1}{2}\right) e^k. \tag{33}$$

Thus, beer acidity P when $t_0 \leqslant t < \infty$ can be determined by the equation:

$$P = \frac{2\,K/K_1}{1 + \left(2\,\frac{K}{K_1} \cdot \frac{1}{P_0} - 1\right) \exp^{[-K(t-t_0)]}}. \tag{34}$$

Now

$$\lim_{t \to \infty} P = 2\,\frac{K}{K_1}. \tag{35}$$

This limit is designated as P_∞. Then Eq. (34) assumes the form:

$$P = \frac{P_\infty}{1 + \left(\frac{P_\infty}{P_0} - 1\right) \exp^{[-K(t-t_0)]}} \tag{36}$$

where P is the beer acidity at time t, $t_0 \leqslant t < +\infty$,
P_0 is the beer acidity when $t = t_0$, and
P_∞ is the beer acidity when $t \to \infty$. P_∞ is not dependent on the initial conditions and is determined by the nature of the bacteria and by the properties of the medium.
Batch alcohol fermentation results in ripe beer, which then undergoes distillation.
The situation is rather different in continuous alcohol fermentation. After a certain time, the alcohol and yeast cell concentrations in each vessel remain practically unchanged until the vessel is emptied, cleaned, and sterilized.
In order to provide efficient continuous fermentation, it is necessary to conduct preventive sterilization not only of the fermentation batteries but also of malting and enzyme house equipment, saccharifying equipment, pumps used to transfer malt extract and wort, wort collectors, refrigerators, tubing, and other equipment.

1.4 Preventive Sterilization of Equipment

Fermentation battery sterilization is conducted in such a way as to reduce contamination without discontinuing the process. Beer from the first head fermenter is pumped into the second, which then becomes the head fermenter charged directly with fresh wort. The empty fermenter is thoroughly washed and treated by steam at 95 °C for 40 min. After cooling, the fermenter is again filled up with yeast and fresh wort. At the same time, the contents of the second vessel are pumped into the third. In this way, the rest of the vessels undergo sterilization in turn. During sterilization of the last fermenter, the ripe beer from the previous vessel is continuously passed to distillation.

This procedure allows for regular sterilization of the equipment while maintaining continuous fermentation in a single battery [23–25].

Apart from batch processes, two sterilization schedules for batteries have been evolved and introduced in the alcohol industry, one series for semicontinuous cyclic fermentation and another for continuous fermentation. These are illustrated, together with batch fermentation for comparison, in Fig. 5. The main advantage of continuous processes is regular preventive sterilization of the vessels to ensure cleanliness of fermentation. The situation is different in cyclic fermentation, during which the sterilization

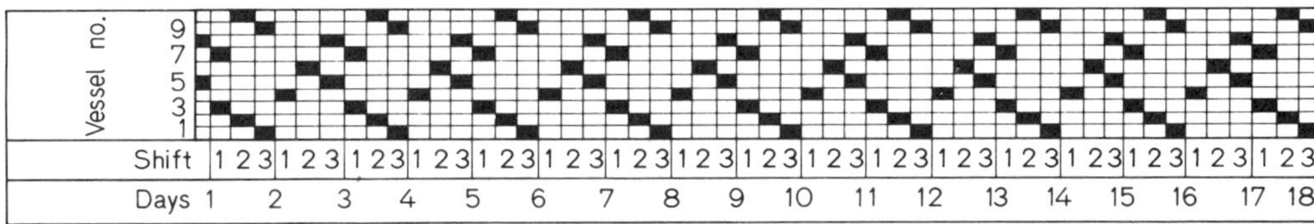

A Batch fermentation

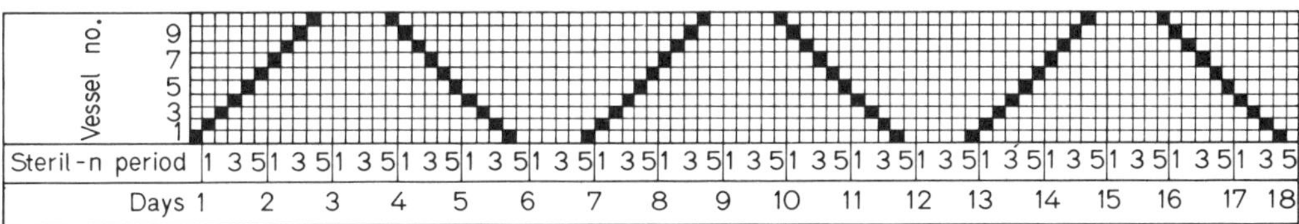

B Cyclic fermentation

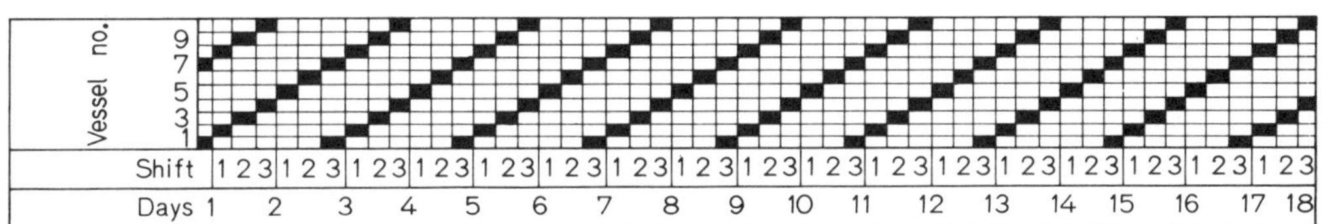

C Continuous fermentation

Fig. 5 A–C. Schedule of vessel sterilization

intervals for head fermenters are long in comparison with intermediate and final ones. Consequently, the ripe beer at the end of the process tends to become sour. To prevent this, beer from cyclic fermentation operations is directed to distillation in an incompletely fermented state.

1.5 Kinetics of Starch Conversion by Enzymes

It is necessary to consider another feature of continuous alcoholic fermentation of starchy media. It is known that the yeast *Saccharomyces cerevisiae* used in alcohol fermentation does not contain enzymatic systems capable of saccharifying starch; this is achieved by using malted grain, fungal enzymes, or bacterial cultures. Before the beginning of fermentation, only partial saccharification of the starchy media occurs (from 25%–30% on the reducing substances basis), and the process goes on in parallel with fermentation for two to three days.

The kinetics of continuous alcohol production on starchy media can be described by the equation of substrate balance [15]:

$$\frac{dS}{dt} = DS_0^1 + D\lambda Z_0 - DS^1 - \frac{1}{Y}\ \mu X \tag{37}$$

where S_0^1 and S^1 = sugars fermentable by yeast contained in the flowing medium,
Z_0 = starch and dextrins of initial medium (calculated as glucose),
λ = coefficient, characterizing rate of starch conversion by enzymes,
D = dilution rate,
X = biomass concentration,
μ = specific growth rate of microorganisms, and
Y = biomass yield constant.

The coefficient λ in Eq. (37) is equal to [15]:

$$\lambda = 1 - e^{Ct}, \tag{38}$$

where C ist rate constant for starch hydrolysis in a monomolecular reaction.
Then Eq. (37) assumes the following form:

$$\frac{DS^1}{dt} = DS_0^1 + DZ_0\,(1\ e^{-Ct}) - DS^1 - \frac{1}{y}\mu X. \tag{39}$$

Analogous equations have been deduced for multistage systems, in which starch hydrolysis occurs in single- and dual flow systems [16], as described later.

2. Continuous Industrial Processes

A number of semicontinuous and continuous flow processes have been developed. The semicontinuous processes comprise the so-called outflow-inflow and overflow processes and also a wide range of battery and cyclic fermentation variants. Characteristic of these processes is the continuous feed liquid flow after addition of a nutritive wort and of a seeding culture, such as *Sacch. cerevisiae r. XII.* The treated medium flows by gravity or is pumped as a seeding culture from the first vessel into the second, while the first continues fermentation. Then, in succession the second fermenter is filled up with the nutritive wort and left for fermentation. Thereafter, the third vessel is filled with the seeding culture, and a prolonged nutritive inflow charge is secured. This process continues until all the vessels are charged.

In contrast to outflow-inflow fermentation, the batterycycle technique (Fig. 6) is based on a steady nutritive inflow through interconnected tubes from the moment of the

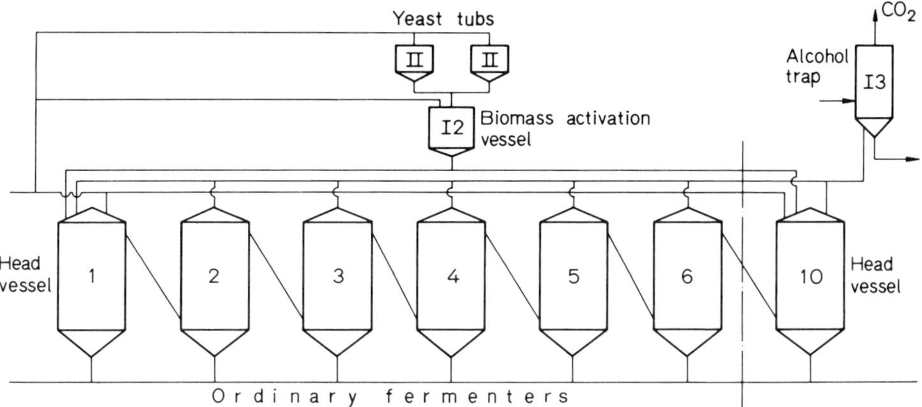

Fig. 6. Scheme of cyclic fermentation

first fermenter charge-up to the filling of the whole battery, which is composed of 6–10 vessels. The after-fermentation period lasts 24–30 h from the filling of the last fermenter. The inflow and seeding culture are then directed to previously discharged and sterilized vessels 6 and 10, and feed-back of the battery is carried out. Subsequently, the procedure is repeated and the ripe culture liquid (beer) is let out of every vessel. The main shortcoming of the process is the varied fermentation medium residence time in each battery of fermenters.

The highest (120 h) and the lowest (24 h) durations of residence time are obtained for the head and back fermenters respectively. The rest are characterized by intermediate values (Fig. 5B). The increase in residence time leads to enhancement of beer acidity and loss of carbohydrates with consequent loss of the final product (alcohol). The curtailment of fermentation time is accompanied by losses through unfermented carbohydrates. These circumstances have led to increased interest in continuous-flow fermentation processes.

2.1 Continuous Flow Process

This process is based on the idea that with continuous inflow of nutrient medium and continuous ripe culture liquid (beer), normal fermentation indices are secured. The fermentation techniques include single and multistage systems. In the USSR, multistage systems composed of 6–12 battery fermenters are usually used (Fig. 7). They can be used in the production of alcohol from miscellaneous crude materials (about 70 plants), champagne-making, primary and secondary fruit wine-making, brewing, fodder yeast production, etc.

An important feature of continuous fermentation is the propagation of seed culture, which occurs in special seed fermenters 1, 2, 3, from which the culture moves by gravity or is pumped into the sterilized fermenter. Such sterilization is carried out from 36 to 48 h. The battery starts operating from the moment the seed culture (e.g. *Sacch. cere-*

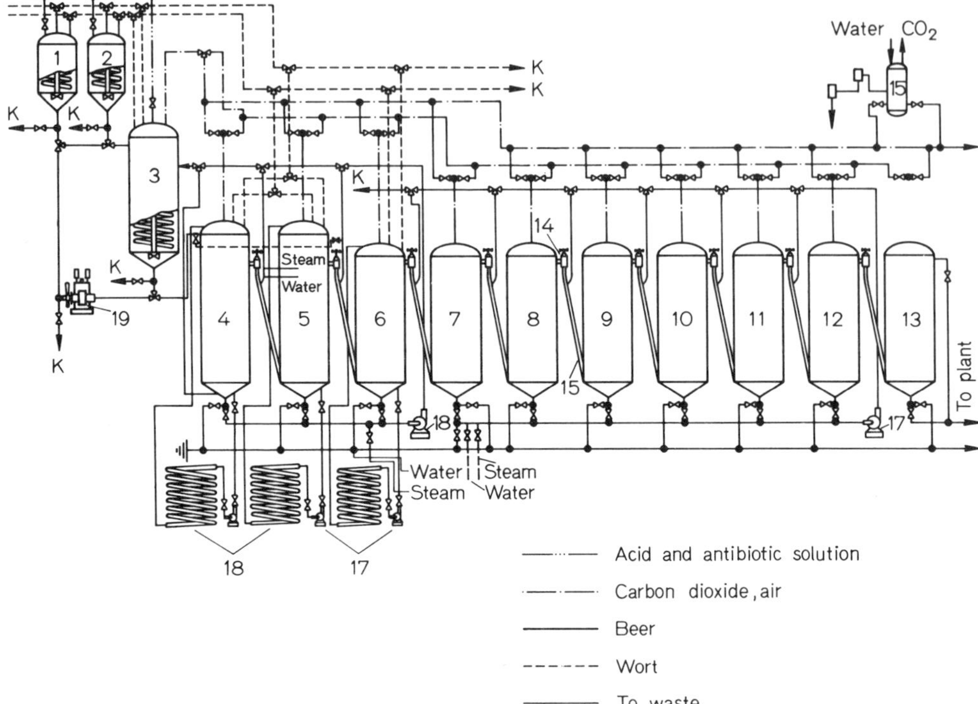

Fig. 7. Continuous fermentation flow-sheet in dynamic cultivation process. 1, 2 yeast tubs; 3 biomass activation vessel; 4–13 fermenters; 14 slide valves; 15 tubing; 16 alcohol trap; 17 centrifugal pumps; 18 heat exchangers; 19 piston pump

visiae r.XII) along with nutrient medium is brought into the head fermenter 4. When the first vessel is full, the fermentation medium flows through interconnecting tubes into vessel 5 then into vessels 6, 7, 8, 9, 10, 11, 12, and 13 successively. Continuous nutrient medium influx and withdrawal of beer that is directed to further treatment are thereby maintained in the battery.

Since the process is not operated aseptically, every 2–3 days successive sterilization of all the battery vessels, tubing, pumps, and other devices is carried out. It is essential to maintain the continuity of nutrient influx and beer withdrawal during the sterilization period, as described later.

In batch processes, sterilization is conducted at certain intervals, but it is not successive. As far as the battery and cyclic processes are concerned, their shortcomings are due to the nature of the processes. The residence times of fermentation medium in the head and back battery vessels are equal to 110–120 and 24 h respectively. The rest of the vessels are characterized by intermediate values. In these conditions the increased loss is due on the one hand to unconsumed carbohydrates as a consequence of the reduction of fermentation time, and on the other hand to acidification in vessels with an extended fermentation time. The drawbacks of the battery fermentation process are clearly seen from the data presented in Table 1. Only initially acid media (pH = 2.5–3.5) are not influenced by the residence time in the reaction zone, since they are not susceptible to microbial contamination and acidification. Weakly acid media (pH = 4.6–6.5) cannot resist infection, and the slightest prolongation of their residence time in the reactors is fraught with the danger of souring and losses, even to the extent of stopping the fermentation. That is why pasteurization or sterilization of media before fermentation is indispensable when conducting continuous processes. But their action is limited to 2–3 days or (at low temperature) 15–20 days, after which sterility fails for various reasons. When this happens, measures should be taken to retain or to restore sterility. Only nonaseptic fermentation can go on for months without complete medium change and repeated input of seeding culture as, for example, continuous cultivation of the fodder yeast *Candida tropicalis* and others.

2.2 Fundamentals of Continuous Culture

The main principles of continuous fermentation have been developed on the basis of 20-years experience of plant operation and extensive research work.

As can be seen from Fig. 8, continuous fermentation has a higher reaction rate than do batch processes. In consequence, the same wort in the continuous process is fermented for 55 h and in the batch process for 73 h the corresponding reaction rate constants equal 0.0512 and 0.0386 h^{-1}. During the first 24 h of continuous fermentation, the greater part (78%) of the sugars is fermented, in the batch process only 42%. This can be explained by the difference in the quantities of yeast present: in the continuous process, the reaction starts in a medium with 90–100 million yeast cells per ml, in the batch process with 12–15 million per ml.

The operation of the fermentation battery is critically dependent on the conditions in the head vessel. Therefore, it must be provided with sufficient seed culture, nutrient

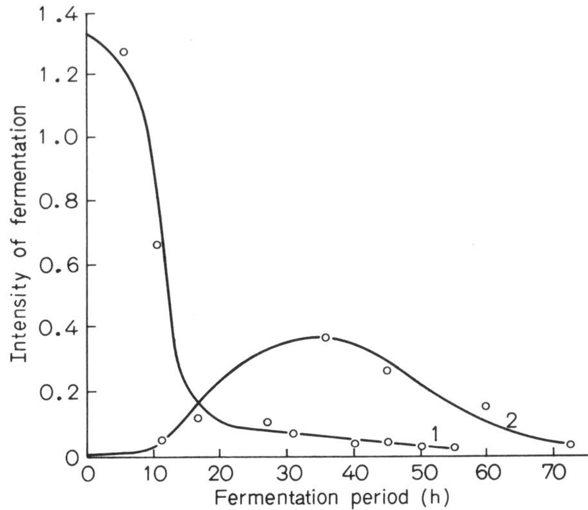

Fig. 8. The curves of sugar fermentation in continuous (1) and batch (2) processes (fermentation intensity is determined by average hourly decrease in dried matter content)

Table 1. Characteristics of fermented media

Fermented medium	Titratable acidity	pH		Solids concentration	Main carbohydrates in medium	Wort sterility		Duration of fermentation
		Initial	Final			Yeast production	Alcoholic fermentation	
1	2	3		4	5	6	7	8
Saccharified potato and grain wort in alcohol production	0.2–0.3	5.5	4.5	16	Glucose, maltose, dextrins	Pasteurized (80 °C)	Sterilized (140 °C)	Batch, 72 h continuous, 60 h
Molasses syrup for alcohol	In dual-flow scheme: 0.6–0.7 In single-flow scheme: 0.3–0.5	4.3 4.3	4.0 4.0	22 22	Saccharose, raffinose (invert sugar)	Pasteurized (80 °C)	Nonsterile, pasteurized (in case of molasses affected by contaminating microorganisms at 85 °C)	24 h
Molasses syrup for yeast production	0.2–0.6	4.5	5.0	2.5–4.0	Saccharose, raffinose	Pasteurized (85 °C)	Pasteurized (85 °C)	24 h
Molasses and flour wort for acetone-butanol production	0.3–0.5	4.5	6.2	5	Starch, invert sugar, glucose, dextrines	Sterilized (130 °C)	Sterilized (140 °C)	72 h
Sulphite liquor for alcohol fermentation	0.4–0.5	5.5	5.0	2.4–2.8	Glucose, mannose, galactose	Sterilized	Sterilized	8 h

Table 1 (continued)

1	2	3		4	5	6	7	8
Wood hydrolysate for alcohol fermentation	0.5–0.7	4.2	4.0	3.0–3.6	Glucose, mannose, galactose, xylose	Sterilized	Sterilized	8 h
Beer wort	0.45	5.65	4.0	11–20	Maltose, dextrins	Pasteurized (100 °C)	Pasteurized (100 °C)	Main fermentation 168 h
Must	2.0	2.5–3.5	2.2–3.4	18–20	Glucose, saccharose	Pasteurized (80 °C)	Pasteurized (80 °C)	Main fermentation 120–192 h
Distillers' spent molasses for fodder yeast growth	0.5–0.7	4.1	4.0	7.0–8.5 not diluted	Traces of raffinose, melibiose, invert sugar	Pasteurized (90 °C)	Pasteurized (90 °C)	12 h
Distillers' spent potato and grains for fodder yeast growth	0.45–0.6	4.8	4.2	6.7–8.0 (grain) 4.1 (potato)	Traces of xylose, arabinose, dextrins	Pasteurized (86 °C)	Pasteurized (86 °C)	8 h

influx at optimal dilution rate, and frequent scheduled sterilization cycles. The violation of these conditions causes failure in normal battery operation, despite the fact that measures are undertaken to rectify the procedure in other fermenters.

The enhancement of acidity in the fermentation media depends on the bacterial population density (Table 2). The rate of increase of acidity has been studied in two fermentation batteries with continuous influx, with yeast cell numbers amounting to 41 million per ml in the first battery and 74 million per ml in the second. Greater acidification was observed with the lower cells number (after 120 h), but at 74 million per ml the increased acidity was observed after 240 h, i.e., 120 h later. In continuous fermentation this time is quite sufficient for the complete conversion of the total wort carbohydrate content into alcohol without losses.

Prolonged experimental testing on a commercial scale has shown that the increase of seed yeast concentration increases the intensity and efficiency of sugar fermentation and hampers the growth of beer acidity (Table 3). By combining the beer samples (100 ml each) that leave the last vessel and keeping the mixture in a thermostat at the same temperature as the continuous culture, a large quantity of lactic acid bacteria and an earlier development of beer acidity was observed when the fermentation started with a low yeast content [27]. Although the beer acidity was initially constant, it appeared to increase sharply after 96 h fermentation and was observed in the head vessel about 48 h later than in the tail one. This can be accounted for by the fact that the head vessel is continuously charged with fresh wort having a low acidity value, and the volume fraction of initial liquid is less than in the following battery vessels.

As the reaction proceeds, the bacterial population increases progressively [20]. The beer residence time increases with the increase in the number of vessels in the battery (Table 4). It is also longer in multistage systems than in single-stage and batch systems. It has been ascertained that in continuous processes the duration of fermentation and residence time in a system are not the same thing.

Irregular flow of medium in a continuous reactor makes it possible for foreign microflora to spread through the battery and propagate directly with the number of battery cycles. The acid-forming microflora along with yeast cells are dispersed throughout the system and are withdrawn from the tail fermenters. However, the latter account for the largest accumulation of microflora, which increases with increasing beer retention time. Since the tail vessels are fed with almost fully fermented beer, acidity decreases in the direction opposite to that of the flow.

If the medium activity in the head fermenter exceeds 0.35 °C the carbohydrate conversion rate slows down, and in 20–30 h the fermentation may cease because of enzyme inactivation. Consequently, the beer leaves the battery containing a large quantity of unfermented carbohydrates. Because of the continuous influx, the beer retention time in the battery increases indefinitely; as a consequence, proliferation of microflora approaches a maximum level. Growth is hampered mainly by the accumulating metabolic products, which inhibit and weaken the yeast and the activity of enzyme participating in dextrin saccharification and sugar fermentation.

Table 2. The relationship between acidity increase and yeast concentration in continuous alcoholic fermentation

Fermentation time (h)	First battery					Second battery				
	Head fermenter			Acidity°			Head fermenter		Acidity°	
	Wort concentration % saccharometer	Acidity°	Yeast cell number (10^6/ml)	Beer on leaving battery	Mixed samples of beer	Wort concentration % saccharometer	Acidity°	Yeast cells number (10^6/ml)	Beer on leaving battery	Mixed samples of beer
0	18.3	0.15	–	–	–	–	–	–	–	–
24	16.0	0.25	47	0.38	–	11.5	0.30	66	–	–
48	16.0	0.27	48	0.40	–	10.2	0.32	69	0.35	0.35
72	16.5	0.30	37	0.40	0.40	9.5	0.32	68	0.38	0.38
96	15.5	0.35	44	0.40	0.40	9.2	0.35	69	0.40	0.40
120	15.6	0.35	37	0.50	0.50	9.2	0.37	80	0.40	0.40
134	15.0	0.35	36	0.50	0.50	9.0	0.40	84	0.40	0.40
168	15.0	0.45	40	0.55	0.70	8.0	0.40	72	0.40	0.40
192	14.5	0.45	47	0.55	0.75	7.5	0.40	77	0.40	0.40
216	–	0.55	41	0.60	0.80	–	0.40	74	0.40	0.40
240	–	–	–	–	0.90	–	0.45	–	0.45	0.45

Table 3. Dependence of ripe beer quality on the quantity of seeding yeast brought into the head fermenter

Time from the beginning of fermentation (h)	Beer quality after bringing in seeding yeast, % head fermenter volume								
	36			62			78		
	Fermented-out beer % saccharometer		Acidity°	Fermendet-out beer % saccharometer		Acidity°	Fermented-out beer % saccharometer		Acidity°
	Planned	Actual		Planned	Actual		Planned	Actual	
56	0.81	0.8	0.35	0.68	0.5	0.35	0.76	0.3	0.30
60	–	0.5	0.35	–	0.3	0.35	–	0.1	0.35
64	–	0.3	0.40	–	0.2	0.35	–	0.1	0.35
68	–	0.2	0.45	–	0.1	0.35	–	0.1	0.35
72	–	0.2	0.55	–	0.1	0.40	–	0.1	0.35
80	–	0.1	0.65	–	0.0	0.45	–	0.1	0.40
Mean	0.81	0.35	0.46	0.68	0.18	0.38	0.76	0.07	0.36

Table 4. Medium residence times in fermenters

Vessel number	1	2	3	4	5	6
Calculated time	10	20	30	40	50	60
Actual time (h)	60	78	80	100	110	140

2.3 Frequency of Battery Sterilization

The successful operation of continuous cultivation on weakly acidic media is governed by regular sterilization not only of the fermentation battery, but also of the malt and submerged culture (enzyme-producing) equipment, mills, saccharifying devices, pumps transferring malt wort, wort collectors, coolers, tubing, and other equipment.

The main feature of such a treatment is the strictly consecutive sterilization beginning with vessel 4 and ending with vessel 13. The sterilization process comprises pumping the fermenting and nutrient medium from vessel 4 into vessel 5, washing, sterilizing by flowing steam, cooling, seeding the culture charge, and resuming nutrient influx to vessel 4. While fermenter 4 is being refilled, the contents of fermenter 5 are transferred into fermenter 6 by pumping, and the emptied vessel 5 undergoes sterilization. When vessel 4 is full, the medium starts flowing through valve 14 and tube 15 into fermenter 5, which is empty and sterile. From vessel 6 the liquid is transferred into vessel 7 and so on, up to the tail vessel. Figure 5 indicates that only continuous fermentation and consecutive sterilization can secure equal residence times for each vessel. This is the most important factor in controlling contamination in any system.

The frequency of battery sterilization depends on the composition and sterility of the fermentation medium, culture type, temperature, medium reaction, and other fermentation conditions. In alcoholic production in particular, the interval between two consecutive sterilizations is between 48 h and 72 h. With lactocide used as an antiseptic, it is 4 days or more. At intervals of approximately 30 h, the fermented wort from the head vessel 4 is pumped into vessel 5, washed, and set into operation again without steam sterilization. After charging, the beer from this fermenter is directed to consecutive mixing with the content of the rest of the vessels. In this case, reduction of the residence time of the medium in the battery occurs, which is especially important at low yeast cell concentrations.

2.4 Application of Sulfuric Acid or Lactocide as Bactericide for Beer Treatment in the Head Reactor

In order to reduce the pH to 4.2–4.3 the production of alcohol from starchy material, sulphuric acid is added to seeding yeast (prior to introducing it into the head fermenter) until the mash acidity reaches a level of 0.9 ml 1.0 N NaOH per 20 ml.

After establishing the medium acidity at the beginning of the fermentation in the head vessel with continuous fresh wort influx at an acidity of 0.15–0.2°, it is possible to stabilize the acidity in the battery at 0.45° after 24–30 h.

Improvements in sterility have been obtained by applying lactocide instead of pasteurizing and acidifying with sulphuric acid. The lactocide dosages for the yeast wort and head fermenter wort are 150–250 and 50 units per ml respectively. The lactocide-treated medium retians its induced asepticity for 3–4 days of fermentation.

2.5 Influence of Seed Yeast Purity on the Head Vessel and Battery Operation

A principal feature of continuous fermentation lies in the fact that the maximum yeast population density is reached not in the head fermenter, but instead in the second or third fermenter. Since this process operates with nonsterile saccharified wort requiring a systematic change of medium as a preventive measure every 2–3 days, it requires the introduction of a biomass activation vessel with a capacity equal to 50% of that of the head fermenter. In this connection, it is necessary to provide for parallel yeast propagation conducted on pasteurized saccharified wort treated by sulphuric acid or lactocide. The total substitution of green malt by bacterial and fungal enzyme preparations cultivated on sterile media and in sterile conditions makes it possible to considerably reduce the number of preventive sterilizations and to correspondingly reduce the yeast requirement for charging the head fermenter.

2.6 Seed Yeast Preparation

A distinctive feature of continuous fermentation in contrast to classical batch fermentation consists in the use of large quantities of seed yeast at the beginning of battery operation. The procedure brings about an increase in microbial population density in the head fermenter as the nutrient and initial seed yeast volumes become equal only after some time has elapsed from the moment of charging. The increased population density thereby obtained results in longer-preserved sterility of the nutrient and fermentation media and in considerable reduction of lactic acid and other bacterial contamination.

Yeast preparation starts from test tubes with pure cultures on sterile worts in incubator and 2-l-flasks, from which yeast, as a mother culture, is passed into the yeast tube and then into the biomass activation vessel with a capacity equal to 50% of that of the head fermenter. Then the seed yeast is directed to the head fermenter and, together with the fermentation medium, passes through connecting tubes to vessels 4, 5, 6, 7, and so on. Thus, the yeast is cultivated only on worts pasteurized at 70°–80 °C in strictly aseptic conditions. The wort in the biomass activation vessel is also pasteurized at 75–80 °C and made sterile by sulphuric acid or lactocide. The wort entering the head fermenter is treated only with lactocide. This makes natural aseptic fermentation for four days possible. Further research work is being directed to prolonging continuous operation of the battery for up to 10 days without contamination. This will permit an increase in the battery and yeast output to over 30%.

2.7 Replacement of Malt by Microbial Enzyme Preparations

Grain malt contains little or, more often, no enzyme capable of splitting α-1,6-glucan bonds in amylopectin, "limit dextrin", and isomaltose. The application of malt with low enzymatic activity necessitates increased dosage, but alcohol yield is still lower than planned. Additionally, starch losses during malting amount to 16% or more. All this makes it desirable to consider the use of saccharifying materials other than malt. The development of enzyme technology makes the application of bacterial enzyme preparations possible.

For a number of years, the replacement of malt by fungal cultures (*Asp. batatae, Asp. awamori*) and yeast culture (*Endomycopsis bispora* and others) grown submerged on filtered and sterilized spent grains with the addition of flour and nutritive salts has been successfully carried out [28].

Development of the fungal culture comprises two stages. First, the seed material is prepared in a mother culture vessel with a volume equal to 10% of that of the fermenter capacity. Then, the seed culture is transferred to the fermenter and sterile nutrient is added. Growth proceeds at 32 °C for 36–56 h during which time starchy material is added. Experience has proved the need to employ hermetically sealed fermentation equipment for this process.

The grown culture is evaluated by units of activity: *Asp. batatae*-61 contains 1.0–1.2 units per ml AA (Amylolytic activity) of 3.2.1.1 α-1,4-glucan-maltohydrolase [19] and 6.0–7.0 units per ml GA (glucoamylase activity); *Endomycopsis bispora*–40 μmol/ml GA; *Asp. awamori*–2.0 units/ml AA and 30 μmol GA. The enzyme required to convert all the starch during a 72 h process amounts to 1.0 unit AA and 7.0 units GA of starchy matter [11].

The decisive role of glucoamylase (3.2.1.3 α-1,4-glucan glucohydrolase) activity in increasing the enzymatic conversion of starchy material has been demonstrated [19]. It is responsible for hydrolyzing about 30% fo the α-1,6-glucan bonds in a starch molecule, thereby decreasing the remaining content in the beer in a combined saccharification-ethanol system.

The exact determination of enzyme dosage considerably simplifies the improvement of conversion and fermentation by increased enzyme dosage. This can only be achieved by concentrated enzyme preparations and not by malt, fungal, or bacterial culture. Ultrafiltration has proved to be the most suitable method for concentrating enzyme preparations. The retentate with 40–45% solids on a dry basis can be kept for 6–8 months and can be transported over long distances. Hence, the replacement of malt by concentrated enzyme preparations in alcohol production excludes another considerable source of contamination.

Considering the powerful effect of the weakly acidic reaction of the nutrient medium on aseptic fermentation development [1] and summing up the measures worked out to prevent contamination, continuous multistage *Sacch. cerevisiae* cultivation (Fig. 7) appears to offer suitable basis for operation on a commercial scale. Extensive research work should be done in this direction, so that continuous aseptic cultivation may be extended to other microorganisms, nutrient media, and products. However, the experience already accumulated in this field makes it possible for other fermentation industries (brewery, wine-making, bakery) to turn to continuous fermentation.

2.8 Dual-Flow Medium Saccharification in Continuous Culture

Great attention is paid to the smooth-running organization of medium feeding in multistage fermentation systems. For example, the phenomenon observed after dividing saccharified wort into two equal flows, directing them in parallel into the first and second fermenters, distributing the enzyme preparation between them unequally (e.g. 0.75 and 0.25), and combining the flows in the continuous fermentation battery causes an increase in reaction rate where the enzyme dosage is largest, i.e., in the head vessel (Fig. 9).

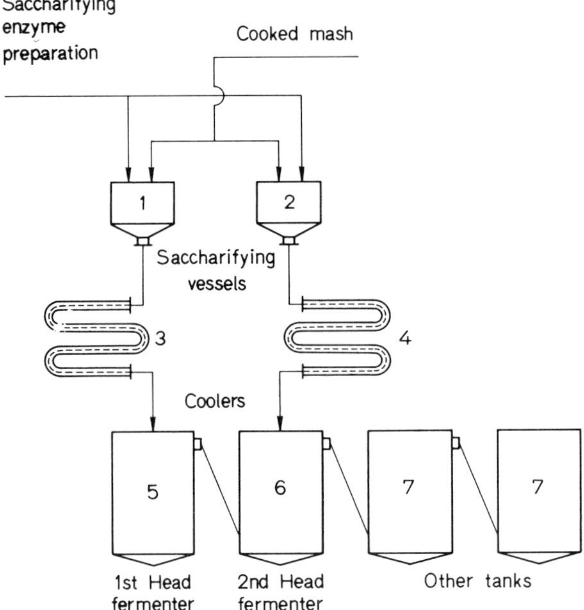

Fig. 9. Dual-flow scheme of saccharification and continuous alcoholic fermentation

	Dosage (units GA/g starch)	Enzyme concentration as % of average for whole dosage of the enzyme added
Flow 1, fermenter 4	0.75 × 7 = 5.25	150
Flow 2, fermenter 5	0.25 × 7 = 1.75	50

The equalization of enzyme concentration takes place as beer flows from the first (head) vessel into the second, then into the third, and so on.

The level of enzyme activity in the first head fermenter is $1\frac{1}{2}$ times that in the other fermenters in the battery [29].

Consequently, starch hydrolysis in the head vessel occurs more rapidly, and conversion into ethanol and other products by yeast is brought about more intensively. This efficient method has been introduced on a commercial scale. It can increase the fermentation rate by 30% and even more (Table 5) [30]. However, dual-flow saccharification and fermentation cannot be applied in single-stage continuous or batch systems.

Table 5. Influence of saccharification system on fermentation characteristics

Fermentation characteristics	Unit of determination	Saccharification system	
		1-flow	2-flow
Starchy material treated	ton	116.0	128.0
Culture dosage	U activity/g starch	8.4	8.4
for saccharifying	% to starch used	88.0	88.0
Wort concentration	%	15.7	15.7
Yeast cell content			
in the 1st fermenter	millions/ml	72.0	84.0
in the 2nd fermenter	millions/ml	72.0	86.0
Alcohol content	% (v)	8.2	8.27
Unfermented carbohydrates	g/100 ml	0.2	0.2
Insoluble starch	g/100 ml	0.03	0.02
Fermentation time	h	55.0	46.0
Alcohol formation	l/day	17,370	22,610
Capacity increase	%		26

3. Further Development in Continuous Culture

3.1 Recirculation of Yeast

According to published data [20, 12], the most significant factor in continuous fermentation is the microbial density. With this in mind, the recirculation of yeast biomass in continuous alcoholic fermentation has been introduced [31].

By employing recirculation, laboratory and pilot plants (Fig. 10) have shown an increase

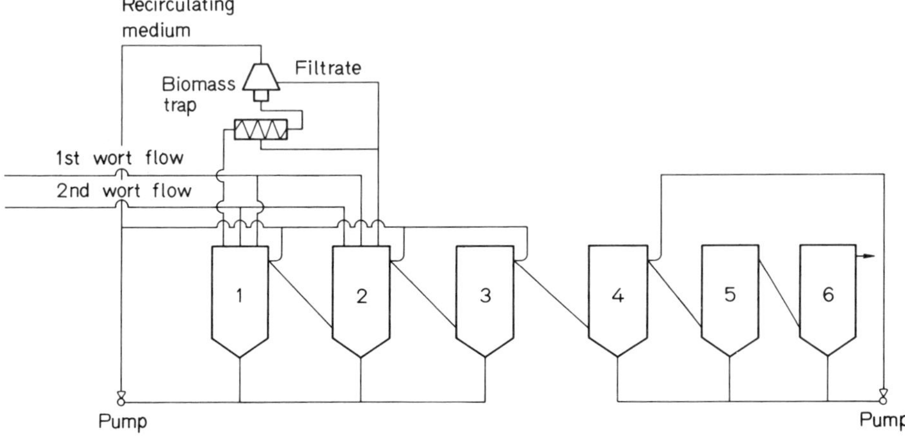

Fig. 10. Continuous fermentation with recirculation and dual-flow saccharification

of dilution rate by 1.7–2.0 times, which corresponds to an increase in fermenter output of 170–200% (Table 6).

Table 6. Enzyme dosage, saccharification system, and influence of yeast recirculation on continuous fermentation time

No. of experiment	Units of enzymatic activity/g starch		Max yeast content (10^6/ml)	Battery feed system	Dilution rate (h^{-1})	Fermentation time (h)
	AC	GC				
1	1.4	5	50.0	single-flow	0.12	68
2	1.4	10	66.0	single-flow	0.14	52
3	1.0	10	100.0	single-flow with recirculation	0.20	35
4	2.0	20	144.0	single-flow with recirculation	0.20	25
5	0.6	9	164	dual-flow with recirculation	$D_1 = 0.21$ $D_2 = 0.10$	22
6	2.0	30	184	dual-flow with recirculation and aminoacid addition	$D_1 = 0.20$ $D_2 = 0.09$	17

The experimental findings show that with increased yeast concentration in the fermentation media and partial recirculation, the growth rate of yeast is increased (μ = from 0.10–0.12 to 0.20–0.30 h^{-1}), with corresponding increase in dilution rate in the battery (D = 0.25–0.31 h^{-1}) [31]. These results are also confirmed by the reaction rate constants, which appear to increase correspondingly from 0.084 h^{-1} in single-flow fermentation to 0.118 h^{-1} in dual-flow saccharification and fermentation and to 0.133–0.161 h^{-1} in dual-flow saccharification with recirculation of yeast biomass. However, alcoholic fermentation with biomass recirculation is accompanied by instability of the microbial population. Nevertheless, the fermentation rate is established from the moment the battery is brought into operation. A certain regulation of fermentation seems to exist, but on a higher level than without recirculation.

As the kinetic characteristics show, the combination of dual-flow saccharification with partial recirculation of microbial biomass ensures the greatest intensification of continuous fermentation. Thus, further research work in this field appears to hold considerable promise.

3.2 Some Critical Remarks with Respect to Procedures and Equipment Used for Separation of Microbial Biomass

According to the information reviewed above, the most urgent task in developing the application of continuous cultivation of microorganisms is to raise considerably the microbial population density throughout the whole reaction zone. Existing methods, unfortunately, do not make this possible. Nevertheless, by applying the principle of closed systems, the task could be accomplished.

For example, the application of packing in the head fermenter is known to promote acetic acid bacteria content in its medium and consequently in the battery output. Similarly, the separation of yeast biomass from culture liquid and its recirculation into the fermenter is practiced in ethyl alcohol production from molasses (CSSR), extracts of ground beet sugar (France), etc.

New Methods of increasing microbial concentration are expected to provide higher specific microbial rate growth (μ) through the corresponding dilution rate increase (D). Our experiments prove that such a connection does exist, but the number of experiments set up in this field is still small, and there is no other available data published on this subject. It is still unknown whether such a relationship will remain valid over a long period of operation and what order of increase of microbial density in the cultivated medium will be attainable.

It also appears necessary to reconsider and reassess the components of the nutrient medium and to further study the mass-transfer processes to secure increased metabolism of cultivated microbes as a basis for increasing their specific growth rates. For instance, owing to lack of amino acids discovered in a number of our experiments, we had to provide another source of amino acids.

There are a number of questions concerned with the constructional features of the equipment used to separate culture liquid into biomass and filtrate in continuous flow. The existing filter presses and the best types of separators and centrifuges currently available are inadequate for this purpose.

The weak points of the above-mentioned equipment lie in the fact that filter presses are provided with insufficiently mechanized systems of cleansing, sterilizing, and feeding filter aids. They usually operate intermittently with incomplete sediment withdrawal in complicated and septic conditions. The filtration of culture liquid through materials with fine pores occurs very slowly. The multicomponent composition of industrial culture liquids also complicates biomass separation.

In principle, the above-mentioned separators are adequate, but they concentrate biomass badly and traumatize live cells by increasing the temperature during separation. Centrifuges are superior in this respect, but filter centrifuges are hardly applicable due to microbial contamination, which remains in the biomass as it does in filter aids used in the culture liquid filtration. Sedimentary centrifuges with intermittent discharge are not appropriate, but still constitute the nearest prototype for developing more sophisticated equipment.

It is desirable that there are increased research and development efforts aimed at securing biomass separation from the fluid in the form of living microorganisms and preserving them for recirculation. Biomass separation, concentration, removal from the separator, and recirculation should be continuous. The provision of sterility in the separation of the culture liquid and preservation of its biological components is important. It is also important to ensure asepticity from the start of cultivation to formation of the product and to maintain aseptic monoculture growth for a minimum of 5–6 days. Considering the experimental results, this task is expected to be accomplished by combining the most effective chemical antiseptic systems with new preventive procedures being developed at present. The practical recommendations elaborated in the course of reserach work are now being tested on the commercial scale.

It is also important that the potentialities of physical and chemical mechanisms acting on culture liquid and its components are closely examined. Ultraviolet irradiation, ultrasonic oscillation, vibrational turbulization, microwave fields, and EM field induction are known from scientific, technical, and patent literature to exert certain effects on various media and materials; their practical application needs investigation.

Further investigations in this field should aim at securing sterility of the media on the one hand and sophistication of the equipment designed for fermentation with concentrated biomass recirculation on the other hand.

The successful fulfillment of these scientific and technical tasks could lead to the doubling or trebling of output per unit volume of vessel capacity. The economic aspect of exploiting this important reserve of productivity is of great significance. Hence, the investigations in this field must be versatile, complex, and accelerated.

Nomenclature

A	liquid initially present in battery of reactors
B	liquid flowing through battery of reactors
B_i	residence time of liquid in i-vessel
$B_{i(r)}$	residence time of liquid in i-vessel at moment τ
C	rate constant of starch hydrolysis in monomolecular reaction
D	dilution rate
°D	acid quantity
I_1, I_2	functional values for the first and second vessels
K	yeast propagation rate (positive constant)
LP_i	Laplace pattern of functional P_i
LR_i	Laplace pattern of functional R_i
N	number of vessels
P	beer acidity at moment t, $t_0 \leqslant t < \infty$
p	Laplace pattern parameter
P_i	new liquid share in i-vessel
$P_i(r)$	new liquid share in i-vessel after r-vessel cycles
P_0	beer acidity under $t = t_0$
P_∞	beer acidity under $t \to \infty$
$P_{j(\tau)}$	part of i-vessel volume (j = 1, 2, ..., i) filled with new liquid at moment τ
r	number of vessel cycles
R_i	old liquid share in i-vessel
$R_i(r)$	old liquid in i-vessel after r-vessel cycles
S_0	new dried material concentration
S_1	dried material concentration in the first vessel
S_2	dried material concentration in the second vessel
ΔS	dried material concentration change during a period of time Δt
S_0^1, S^1	fermentable by yeast sugars contained in circulating medium
T	vessel cycle duration
t	time
v	vessel volume
V	new liquid volume
X	biomass concentration

X_1	number of acid-producing bacteria
X^1	maximum concentration of yeast cells in unchanged medium
$X_i^{(p)}$	$\lim_{t \to \infty} X_i$ = equilibrium concentration of yeast cells in i-vessel
$X_{i-1}^{(p)}$	equilibrium concentration of yeast cells in beer feeding i vessel
Y	biomass yield constant (economic coefficient)
Z	starch and dextrins of primary medium (including glucose).

Subscripts unless otherwise noted

i(i = 1, 2, 3, ..., n)	ordinal number of vessels
∞	value at time = infinity
Δs, Δt, $\Delta t \to 0$	increment approaches zero
max.	maximum value

Greek Letters

φ	resultant sugar fermenting capacity
λ	coefficient, characterizing starch conversion by enzymes
μ	specific growth rate of microorganism
τ	period of time
θ_1	cycle of a fermenter

Literature Cited

1. Braunstein, A. E.: Nomenclatura fermentov. Recomendatsii mejdunarodnogo Sojuza po nomenclature i classificatsii fermentov. Moscwa, 160 (1966).
2. Danilov, K. G./Yarovenko, V. L.: K teorii nepreryvnogo protsesa peremeschenia jidkosti v batareye soobshchayushchikhsya sosudov. Trudy TsHIISP, vypusk 11, 3–14 (1961).
3. Danilov, K. G./Yarovenko V. L.: Sravneniye prosteishikh modificatsii golivnoi chasti brodilnoi batarei. Spirtovaya promyshlennost **4**, 8–14 (1963).
4. Danilov, K. G.: Zapolnenie novoi jidkostyu batarei is sosydov razlichnogo objema. Izv. VUSov. Pishchevaya tekhnologia **3**, 143–139 (1962).
5. Danilov, K. G./Yarovenko V. L.: Prtsess zakisanija zreloi brazki. Trudy TsNIISP, vypusk VIII, 25–28 (1962).
6. Herbert, D.: Continuous Culture of Microorganisms. Some Theoretical Aspects. Continuous Cultivation of Microorganisms. A Symposium. Prague; Publ. House Czechoslov. Acad. Sci, 45 (1958).
7. Herbert, D.: Die Branntweinwirtschaft **100**, 23, 567 (1960).
8. Iyerusalismky, N. D.: Metod protochnogo kultivirovaniya microorganismov. Moscwa; Pishcepromisdat, **9** (1960).
9. Ierusalimsky, N. D.: Microbiologia **30**, 5, 818–825 (1961).
10. Iyerusalimski, N. D.: Upravlyaemii biosyntes. Moscwa; Nauka, **5**, 398 (1966).
11. Kretovich, V. L./Yarovenko, V. L.: Fermentnii preparaty v pishchevoi promyshlennosti. Moscwa; Pishchevaya promyshlennost, 536 (1975).
12. Malek, I./ Fenzl, Z.: Nepreryvnoye kultivirovanie mikroorganismov. Moscwa: Phishevaya promyshlennost. 545 (1968).
13. Monod, J.: Recherches suz la croissance des cultures bacteriennes. Paris: Herman et Cie, **1** (1942).
14. Monod, J.: "Ann. Inst Pasteur" **79**, 390 (1950).

15. Nakhmanovich, B. M./Yarovenko, V. L./ Levchik, L.A.: Kinetika osakharivaniya i nepreryvnogo spirtovogo brozhenia krahmalistykh sred. Fernentnaya i spirtovaya promyshlennost **2**, 9–12 (1972).
16. Nakhanovich, B. M./Yarovenko, V. L.: Kinetika osakharivania i nepreryvnogo spirtovogo brozhenia krakhmalystykh sred v mnogostypenchatych systemach. Fermentnaya i spirtovaya promyshlennost 7, 7–10 (1973).
17. Novick, A./Szilard, L.: Proc. Nat. Acad. Sci. **2**, Washington (1950).
18. Novick A./Szilard L.: Science **112**, 715 (1950).
19. Ustinnikov, B. A./Yarovenko, V., L./Pykhova, S. L.: Proisvodstvo i primenenie glubinnoi kultury plesnevykh grobiv b spirtovoi promyshlennosti. Moscwa: Pishchevaya promyshlennost, **98** (1969).
20. Yarovenko, V. L.: Osnovnyje zukanomernosti nepreryvnogo spirtovogi i atsetono-butilovogo brozhenia. Moscwa: Pishchevaya promyshloennost, **103** (1975).
21. Yarovenko,V. L.: Potochnii metod spirtovogo brozheniya. Moscwa: Pishchepromisdat, **127** (1959).
22. Yarovenko, V. L.: O prodoljitelnosti prebyvaniya zhidkosti v brodilnoi bataree. Moscwa, TsINITI pishcheprom (1963).
23. Yarovenko, V. L.: Sposob zapolneniya batarei brodilnykh chanov syslom pri nepreryvnom protsesse ego zbrazhivaniya. Avtorskoye svidetelstvo 96695. Byulleten isobretenii **1** (1954).
24. Yarovenko, V. L.: Sposob progotovleniya i vneseniya drozhei v brodilnii chany pro nepreryvnom spirtovom brozhenii. Avtorskoye svidetelstov 120203. Byulleten isobretenii **11** (1959).
25. Yarovenko, V. L.: Sterilizatsiya brodilnoi posysy pri nepreryvnom brozhenii. Spirtovaya promyshlennost **3**, 14–17 (1954).
26. Yarovenko, V. L./Nakhmanovich, B. M.: Nepreryvnoye spirtovoye brozheniye krahmalistykh sred s retsirculyatsiei drozhei. Mikrobiologicheskaya promyshlennost 2 (1976).
27. Yarovenko, V. L.: Osnovnyie zakonomernosti nepreryvnogo spirtovogo i atsetono-butilovogo brozhenia. Simposim po nepreryvnosti kultivirovaniya microorganismov. Praga, 227–281 (1962).
28. Yarovenko, V. L.: O factore sterilnosti pri nepreryvnom kultivirovanii microorganismov. Microbiologicheskaya promyshlennost **6** (126), 4 (1975).
29. Yarovenko, V. L./ Vorobyova, L. A./Yarovenko, V. V.: Teoria i practica polucheniya productov brozhenia pri nepreryvnom kultivirovanii microorganismov. Moscwa: Microbiologia **4**, 77–101 (1975).
30. Yarovenko, V. L./Nakhmanovich, B. M.: Dvuchpotochnoye osacharivanie i nepreryvnoye spirtovoye brozhenie krakhmalistykh zatorov. Fermentnaya i spirtovaya promyshlennost **8**, 12–15 (1975).
31. Yarovenko, V. L./Nakhmanovich, B. M./Levchik, L. A./Levchik, A. P.: Sposob proisvodstva spirta is krachmalistogo syriya. Avtoskoe svidetelstvo 384863. Byulleten isobretenii **25** (1973).

Mechanism of Liquid Hydrocarbon Uptake by Microorganisms and Growth Kinetics

Yoshiharu Miura
Department of Biochemical Engineering,
Faculty of Pharmaceutical Sciences, Osaka University,
133-1 Yamadakami, Suita, Osaka, Japan

Contents

1. Introduction

The mechanism of liquid hydrocarbon uptake by microorganisms represents an important problem. The kinetics of microbial growth on hydrocarbon are a basis for industrial development of hydrocarbon processes. The kinetic model for cell growth on hydrocarbon should be based on the mechanism of hydrocarbon uptake.

In microbial cultures with liquid hydrocarbon of low solubility in water, there exist two liquid phases, i.e. the aqueous and the hydrocarbon phase. Numerous discussions have taken place concerning the mechanism of liquid hydrocarbon uptake by microorganisms. In the present review, the uptake mechanisms of liquid hydrocarbon of low solubility in water are first discussed. Then, the growth kinetics of microorganism on liquid hydrocarbon are discussed on the basis of the uptake mechanism of hydrocarbon for microorganisms with low and high affinity for hydrocarbon.

2. Affinity of Microorganisms for Liquid Hydrocarbon

The microbial assimilation of hydrocarbon and the growth on it are considered to be strongly associated with the microbial affinity for hydrocarbon. The adherent characteristics of microorganisms to hydrocarbon and the adhesive force between microorganisms and hydrocarbon are discussed in this section.

2.1 Adherent Characteristics

The microscopical observations of Mimura *et al.* [1] and Blanch and Einsele [2] showed that *Candida petrophilum* and *Candida tropicalis* adhered to the oil phase in cultures with hydrocarbons as a substrate and that the cells and oil droplets formed agglomerates, so-called flocs. Einsele, Schneider and Fiechter [3], in studies with the electron microscope, observed the submicroscopical oil droplets adhering to the cell wall.
The affinity of the cell surface to hydrocarbon was investigated by Kaeppeli and Fiechter [4] using *Candida tropicalis* ATCC 32113. For their experimental conditions, the amount of attached n-hexadecane to the cell surface was about 160 mg $\cdot$ (g dry wt.)$^{-1}$. With the assumptions given in Table 1, a mean thickness of 0.08 μ for the n-hexadecane layer can be calculated. The affinity was basically unaffected by different pH values and temperatures as well as by the chain length of the n-alkene for a pH between 4.5 and 10.0, temperatures from 20 to 50 °C and an n-alkane with a carbon number between 15 and 18. From these results, the interaction between the hydrocarbon and the cell surface is considered to be a passive adsorption that does not involve an enzymatic reaction.
The cells grown on glucose showed a 25% lower adsorption capacity compared to those grown on n-alkane. The cells grown on glucose were more sensitive to the quality of the emulsion. There was a widening of the gap of the adsoprtion capacity with decreasing quality of the emulsion for cells grown on glucose and on hydrocarbons. Those results suggested to Kaeppeli and Fiechter [4] that there are changes in the cell surface related to the growth substrate and that this may partly explain the formation of flocs during the shift from glucose to hydrocarbons reported by Hug, Blanch and Fiechter [5].
The properties of the cell surface of *C. tropicalis* ATCC 32113 were investigated by Kaeppeli [6]. It was shown that there were two peculiarities important for the interaction between the cell surface and the hydrocarbon:

1) porous structure of the surface layer and
2) formation of a mannan-fatty acid-complex during growth on hydrocarbons thereby increasing the lipophility of the surface.

The formation of the mannan-fatty acid-complex was studied during a substrate shift from glucose to hydrocarbons. The decrease of affinity on glucose is considered to be a consequence of a diminished lipophility at the cell surface. The porous structure of the cell surface layer is considered not to be affected by the substrate change. An emulsifying agent which was homologous to the mannan-fatty acid-complex at the cell surface, was isolated from the medium.

Table 1. Assumptions made for the calculation of the thickness of the n-hexadecane layer attached to the cell surface [8]

Mean cell diam.	$4 \cdot 10^{-6}$ m
Cell vol.	$32 \cdot 10^{-18}$ m^3
Cell no./mg dry wt.	$5 \cdot 10^7$
Volume of the attached n-Hexadecane (160 mg)	$2 \cdot 10^{-10}$ m^3
Increase in vol./cell	$4 \cdot 10^{-18}$ m^3

Osumi *et al.* [7] investigated the ultrastructure of the cell surface of *Candida tropicalis* and that of *C. albicans* growing on n-alkanes using field emission scanning electron microscopy. A curious protrusion of 100–200 mm in diameter was observed on the surface of the yeast cells. The protrusions, consisting of small subunits of about 50 nm in diameter, were scattered at a regular distance from each other on the cell surface. The sectioned view of the cells grown on n-alkanes, obtained by transmission electron microscopy, showed the existence of slime-like outgrowths on the periphery of the cell wall. These outgrowths, which were observed as electron-dense layers, developed across the cell wall, thereby producing channels that connected to the cell membrane. The slime-like outgrowth was supposed to correspond to the above-mentioned protrusion on the cell surface.

Using *Candida intermedia* IFO 0761, *Candida tropicalis* ATCC 20336 and *Saccharomyces cerevisiae* Hansen IFO 0305, Miura *et al.* [8] investigated the affinities of the hydrocarbon-utilizable and -unutilizable microorganisms for liquid hydrocarbon by measuring the degree of adhesion between the cell and hydrocarbon. Alanine-precultured cells were used in the investigation of their adherent characteristics to hydrocarbon, since they are considered not to contain the hydrocarbon before the investigation. The results in Table 2 indicate that *C. intermedia* and *C. tropicalis,* which can utilize the hydrocarbon, adhere well to the hydrocarbon but that *S. cerevisiae,* which cannot utilize the hydrocarbon, is not adherent. The number of adsorbed cells per unit surface area of hydrocarbon was not appreciably different for *C. intermedia* and *C. tropicalis.* Whereas clumps were formed by *C. intermedia* and hydrocarbon, *C. tropicalis* did not form clumps with hydrocarbon.

Table 2. Adherent characteristics of microbial cells for hydrocarbon

Strain[a]	Assimilation of n-tetradecane	Adsorption onto n-tetradecane[b]	Formation of clump	Adsorbed cells per unit surface area of n-tetradecane (cells/cm^2)
C. intermedia	+	+	+	3.0×10^7
C. tropicalis	+	+	–	2.2×10^7
S. cerevisiae	–	–	–	0

[a] Alanine-grown cells.
[b] Observed microscopically.

2.2 Adhesive Force

The adhesive force between cells and hydrocarbon was assessed by Miura *et al.* [8], using *Candida intermedia* IFO 0761 and *Candida tropicalis* ATCC 20336 by centrifuging cells adsorbed onto hydrocarbon at different centrifugal forces for 3 minutes and thereafter counting the cells that were desorbed from the hydrocarbon due to the centrifugation. The effect of a surface active agent on the adhesive force was investigated by adding 50 ppm of Tween 20 to the cells that were adsorbed onto hydrocarbon. The results, which indicate that the adhesive force between *C. intermedia* and n-tetradecane is much stronger than that between *C. tropicalis* and n-tetradecane, are shown in Fig. 1. Adsorbed cells of *C. tropicalis* were substantially desorbed from the hydrocarbon phase at a centrifugal force greater than 10000 × g, while about one fourth of the adsorbed cells of *C. intermedia* were desorbed due to a centrifugal force greater than 3 000 × g. The effect of the surface active agent, Tween 20, on the adhesive force between the cell and hydrocarbon is shown in Fig. 1. Figure 1 also shows that all adsorbed

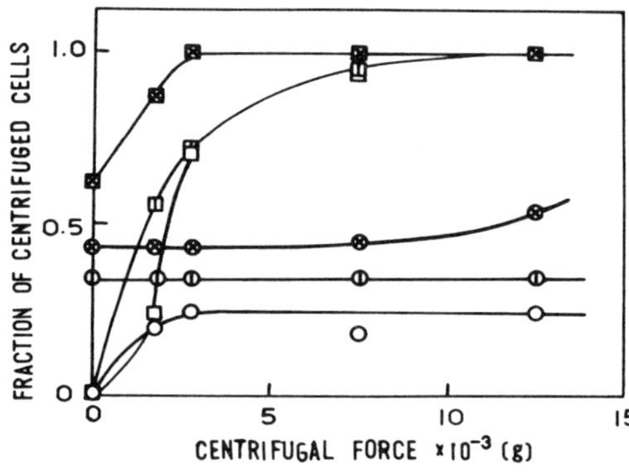

Fig. 1. Degree of Desorption of cells from n-tetradecane under centrifugal forces [8]

- ○ *C. intermedia*
- ⊗ *C. intermedia* with Tween 20
- ⦶ *C. intermedia* with culture filtrate
- □ *C. tropicalis*
- ⊠ *C. tropicalis* with Tween 20
- ◫ *C. tropicalis* with culture filtrate

cells of *C. tropicalis* are desorbed from the hydrocarbon-phase as the result of a centrifugal force greater than 3 000 × g, while only 40 to 50% of the adsorbed cells of *C. intermedia* are desorbed due to the same centrifugal force. In order to investigate whether or not the microorganisms secrete surface active agents which affect the adhesive force between cells and hydrocarbon, the cells cultured in the medium with hydrocarbon were centrifuged, the culture filtrate was then mixed with cells adsorbed onto hydrocarbon and this mixture was centrifuged. The results shown in Fig. 1 indicate that the fraction of cells of *C. tropicalis* that were desorbed is almost the same as that without Tween 20, and the fraction of cells of *C. intermedia* desorbed is lower than that with Tween 20 and higher than that without Tween 20. It is inferred from these results that *C. intermedia* secretes surface active agents but *C. tropicalis* does not.

From these results it is assumed that the affinities of the hydrocarbon-utilizable microorganisms for hydrocarbon are different, depending on the kinds of microorganisms involved.

3. Pathway for Liquid Hydrocarbon Uptake

In hydrocarbon fermentation, the three different possible pathways considered for liquid hydrocarbon uptake are as follows:

1) uptake of dissolved hydrocarbons in aqueous phase,
2) direct contact of the cells with submicron oil droplets, i.e. the accommodated oil droplets proposed by Aiba *et al.* [9] and
3) direct contact of the cells with large oil drops.

It was reported by Erdstieck and Rietema [10] that the uptake of dissolved hydrocarbon should not be neglected for the microbial growth when undecane and dodecane are used as a carbon substrate. Yoshida, Yamane and Yagi [11, 12] also showed that the utilization of dissolved hydrocarbons might contribute to the microbial growth when decane and shorter chain hydrocarbons are used as a carbon source. Chakravarty *et al.* [13] observed that the microbial cells did not stick to the hydrocarbon particles during growth on emulsified solid parafins. Based on this observation, they postulated a model for microbial growth on solid hydrocarbons, employing the assumption that growth occurred on dissolved substrate and that the dissolution of the substrate was helped by a metabolite produced by the cells. This model was able to explain the observed growth characteristics.

On the other hand, Johnson [14] denied the possibility of microbial uptake of hydrocarbons dissolved in the aqueous phase, in view of the very low solubilities in water of n-paraffins heavier than C_{10} and proposed that, in fact, the microorganisms took up liquid hydrocarbons by direct contact with hydrocarbon drops. Aiba *et al.* [9] and Aiba and Haung [15] reported that their experiments with a strain of *Candida guilliermondii* yielded a ratio for the amount of n-alkanes which dissolved in the aqueous phase during fermentation to the total amount of n-alkanes consumed by the cells of approximately 0.9%. They therefore proposed that the uptake of dissolved hydrocarbon can be neglected, when comparatively longer chain hydrocarbons such as hexadecane are used as a substrate.

Dunn [16] discussed the shapes of the growth rate curves for both the "interfacial kinetic model", which assumed direct contact between hydrocarbon drops and microorganisms, and the "homogeneous kinetic model", which assumed the microbial uptake of hydrocarbon dissolved in the aqueous phase, but he did not show which of the two models was valid in the usual case. Several researchers (Erickson *et al.* [17]; Prokop *et al.* [18]; Wang and Ochoa [19]; Katinger [20]) reported that the yeast cells grew mainly at the interface between oil and water and that the growth was limited by the interfacial area. Erickson *et al.* [17, 21–23] considered all the conceivable mechanisms of hydrocarbon uptake, including that of direct contact and the dissolution in the aqueous phase, and proposed various mathematical models. However, their theories were not verified experimentally.

Katinger [20] reported that the interfacial area per unit oil fraction increased as cultivation proceeded. Wang and Ochoa [19] and Prokop *et al.* [18] showed, respectively, that the cells caused an increase of the interfacial area. Mimura, Watanabe and Takeda [1] pointed out the importance of the formation of flocs, which consisted of cells, oil drops and air bubbles, using a strain of *Candida petrophilum.* This yeast seems to have a high affinity for hydrocarbon.

It has been suggested by Aiba *et al.* [9, 15], Moo-Young, Shimizu and Whitworth [24, 25], Goma *et al.* [26], Yoshida *et al.* [27, 28], Hisatsuka *et al.* [29] and Chakravarty *et al.* [30] that the liquid hydrocarbons available to the cells were mainly submicron droplets. Considerable evidences have been presented supporting this proposition, as shown in Section 5. The microorganisms used by these researchers seem to have low affinities for hydrocarbons.

However, Einsele *et al.* [31] reported that there might be free cells, cells which might adsorb submicron droplets, and cells attached onto large oil drops, in the hydrocarbon fermentation. Bakhuis and Bos [32] investigated the relation between the growth rate of *Candida tropicalis* and the size of oil droplets and showed that the growth rate was the lowest for the oil droplets size of 20 to 25 μ and that it was about twenty times higher than this minimum for the droplet size of 16 μ but only about five times higher at the droplet size of 29 μ.

Recently, Nakahara *et al.* [33] showed through microscopic observation and the results of batch cultivation in the tower system, that many cells attached to large oil drops formed flocs as has been reported by Mimura *et al.* [1] and Einsele *et al.* [31]. However, it was also observed that a significant fraction of the cells were free from large oil drops. About 30% of the cells in the culture broth precipitated during centrifugation. This suggests that more than one-half of the cells may be attached to large oil drops. Therefore, both processes, i.e. direct contact of cells with large drops and utilization of submicron droplets, may be important for hydrocarbon uptake by microorganisms.

Nakahara *et al.* [33] also showed that the interfacial tension between oil and water decreased sharply as cultivation proceeds and as the size of the oil drop decreases. They investigated the effect of various fatty acids as well as cell density on the interfacial tension, demonstrating that long chain dioic acids and cells reduced considerably the interfacial tension but that palmitic acid had only a slight effect on it and that the spreading coefficient increased with decreasing interfacial tension. According to these results, they suggested that in batch fermentation the relative contributions to the microbial growth of large oil drops and submicron droplets may change as the cultivation proceeds because of the decrease in drop size and due to an increase of the pseudosolubility (presented by Goma *et al.* [26]) as the result of substantial changes of the interfacial tension. Shah *et al.* [34] came to the same conclusion.

The author *et al.* [8] investigated the pathway for liquid hydrocarbon uptake, using several microorganisms with different affinity for hydrocarbon, and showed that the pathway for liquid hydrocarbon uptake depends upon the affinity of the microorganisms for hydrocarbon. These results are presented below. In this work, the drop-form hydrocarbon refers to hydrocarbon drops of larger size than the cell and the accommodation-form hydrocarbon refers to hydrocarbon droplets of smaller size than the cell. The total quantity of drop- and accomodation-form hydrocarbons utilized, was compared with that of the dissolved hydrocarbon by measuring the quantity of ^{14}C-n-decane uptake by *Candida lipolytica* NRRL Y-6795, precultured on alanine. Further, in order to ascertain ^{14}C-n-decane uptake by the microorganism, the quantity of $^{14}C\text{-}CO_2$ produced by the microorganism was measured.

Utilization of the drop- and accommodation-form ^{14}C-n-decanes was much higher than that of dissolved ^{14}C-n-decane, the latter being negligible as shown in Fig. 2 although sufficient dissolved ^{14}C-n-decane remained in the medium after this measurement. The quantity of ^{14}C-CO_2, produced by *C. lipolytica* from the drop- and accommodation-form ^{14}C-n-decanes, was also much higher than that from dissolved ^{14}C-n-decane, the latter again being negligible as shown in Fig. 3.

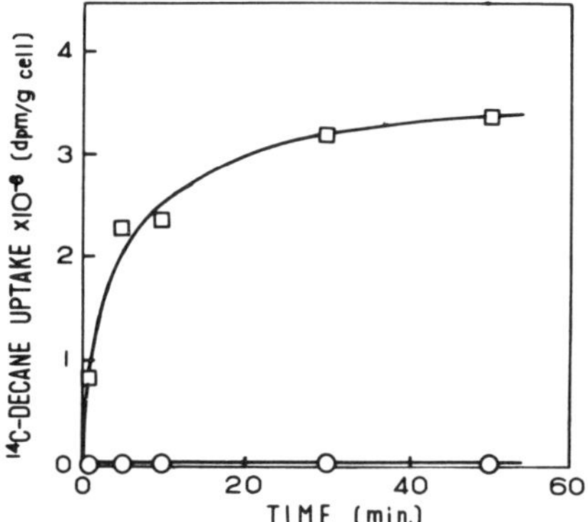

Fig. 2. Comparison between the ^{14}C-n-decanes uptake by *C. lipolytica* on drop- and accommodation-form ^{14}C-n-decanes and that of dissolved ^{14}C-n-decane [8]
□ drop- and accommodation-form ^{14}C-n-decanes
○ dissolved ^{14}C-n-decane
cell concentration: 0.02 g/l

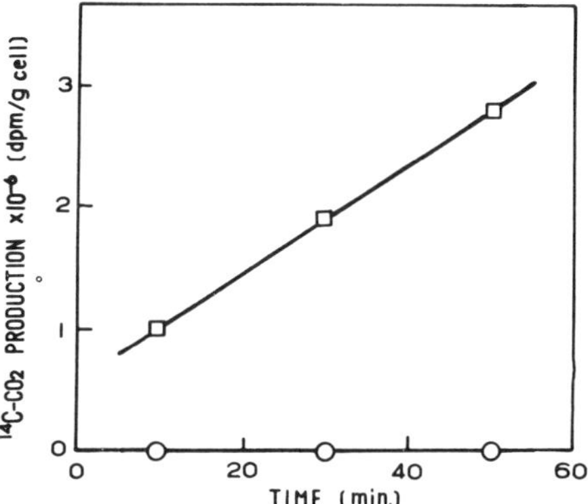

Fig. 3. Comparison between the ^{14}C-CO_2 produced by *C. lipolytica* on drop- and accommodation-form ^{14}C-n-decanes and that of dissolved ^{14}C-n-decane [8]
□ drop- and accommodation-form ^{14}C-n-decanes
○ dissolved ^{14}C-n-decane
cell concentration: 1 g/l

From the results mentioned above it is concluded that the quantity of dissolved hydrocarbon utilized is negligible compared with the quantity of drop- and accommodation-form hydrocarbons utilized, when comparatively longer chain hydrocarbons such as decane are used as a substrate.

In order to compare the uptake rate of accommodation-form n-tetradecane with that of drop-form n-tetradecane, the rates of oxygen uptake were measured on drop-form and accommodation-form n-tetradecanes, since the oxygen uptake rate was proportional to the specific growth rate as shown in Fig. 4. The microorganisms used were *Candida intermedia* IFO 0761, with high affinity for hydrocarbon and *Candida tropicalis* ATCC 20336, with low affinity, both were mentioned in Section 2. The rates of oxygen uptake on accommodation-form n-tetradecane were measured using various substrate concentrations for alanine-precultured *C. intermedia* and *C. tropicalis* at 0.066 and 0.063 $g \cdot l^{-1}$, respectively. The results are shown in Figs. 5 and 6 and indicate that the maximum rates of oxygen uptake on accommodation-form n-tetradecane, $Q_{O_2}(AH)_{max}$, are as follows:

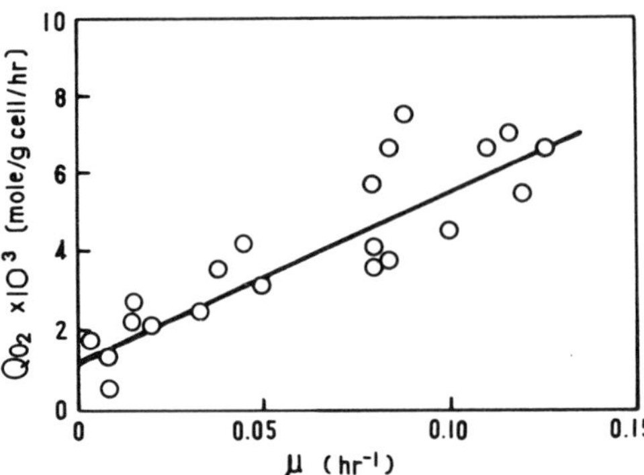

Fig. 4. Relationship between oxygen uptake rate, Q_{O_2}, and specific growth rate, μ, for *C. C. intermedia* [8]

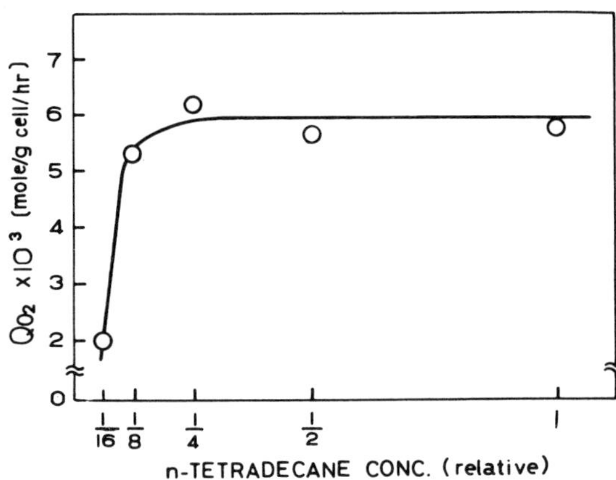

Fig. 5. Oxygen uptake rate, Q_{O_2}, on accommodation-form n-tetradecane for *C. intermedia* [8]
Unity of relative n-tetradecane concentration: 5000 ppm
Cell concentration: 0.066 g/l

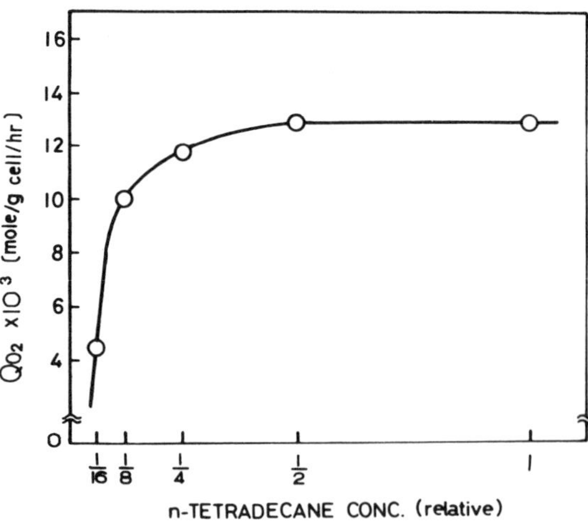

Fig. 6. Oxygen uptake rate, Q_{O_2}, on accommodation-form n-tetradecane for *C. tropicalis* [8]
Unity of relative n-tetradecane concentration: 5000 ppm
Cell concentration: 0.063 g/l

for *C. intermedia*

$$Q_{O_2}\,(AH)_{max} = 6 \times 10^{-3}\ (\text{mole } O_2) \cdot (\text{g cell})^{-1} \cdot h^{-1}$$

and for *C. tropicalis*

$$Q_{O_2}\,(AH)_{max} = 1.3 \times 10^{-2}\ (\text{mole } O_2) \cdot (\text{g cell})^{-1} \cdot h^{-1}.$$

The rates of oxygen uptake at various concentrations of alanine-precultured cells were measured with a definite initial concentration of accommodation-form n-tetradecane, 5000 ppm. These rates of oxygen uptake are shown in Fig. 7 for *C. intermedia* and in Fig. 8 for *C. tropicalis.* According to the results shown in Figs. 7 and 8, the maximum rates of oxygen uptake are as follows:

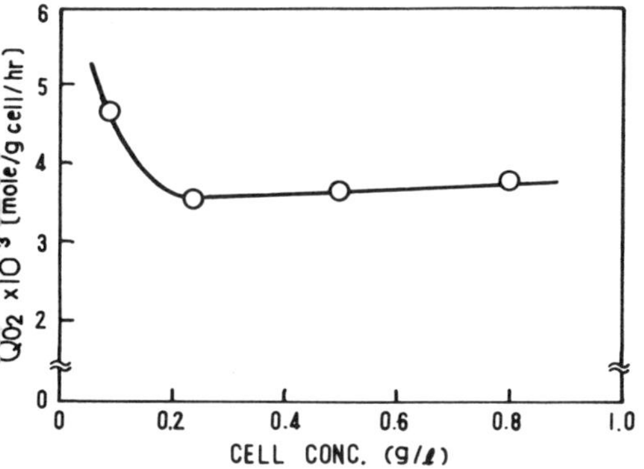

Fig. 7. Oxygen uptake rate on accommodation-form n-tetradecane for *C. intermedia* [8] *media*
n-tetradecane concentration: 5000 ppm

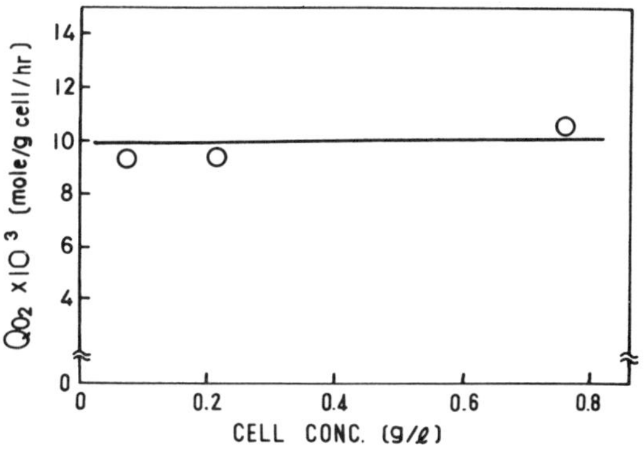

Fig. 8. Oxygen uptake rate on accommodation-form n-tetradecane for *C. tropicalis* [8]
n-tetradecane concentration: 5000 ppm

for *C. intermedia*

$$Q_{O_2}\,(AH)_{max} > 4.7 \times 10^{-3}\ (\text{mole } O_2) \cdot (\text{g cell})^{-1} \cdot h^{-1}$$

and for *C. tropicalis*

$$Q_{O_2}\,(AH)_{max} = 1 \times 10^{-2}\ (\text{mole } O_2) \cdot (\text{g cell})^{-1} \cdot h^{-1}.$$

$Q_{O_2}\,(AH)_{max}$, obtained by varying the n-tetradecane concentration, was approximately equal to that obtained by varying the cell concentration.

In order to assess the uptake rates of drop-form hydrocarbon, the oxygen uptake rates of alanine-grown cells on drop-form n-tetradecane were measured for various cell concentrations. Figure 9 shows the oxygen uptake rate for *C. intermedia* and Fig. 10 shows the rate for *C. tropicalis.* The maximum rates of oxygen uptake on drop-form hydrocarbon, $Q_{O_2}(DH)_{max}$, were as follows:

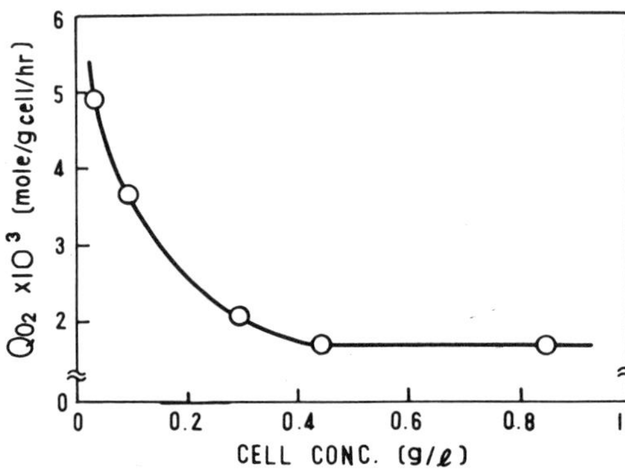

Fig. 9. Oxygen uptake rate on drop-form n-tetradecane for alanine-grown *C. intermedia* [8]
n-tetradecane concentration: 10^5 ppm

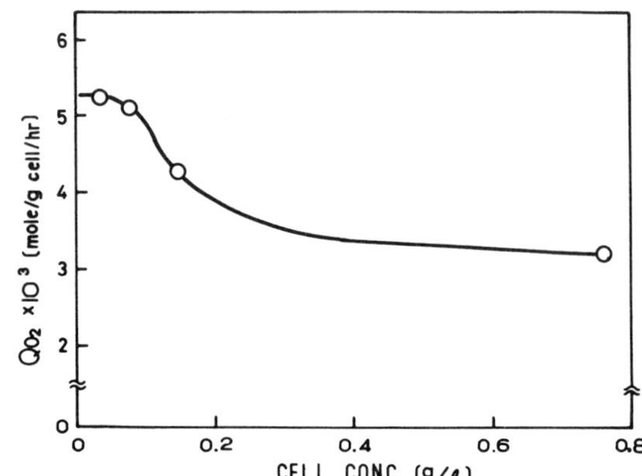

Fig. 10. Oxygen uptake rate on drop-form n-tetradecane for alanine-grown *C. tropicalis* [8]
n-tetradecane concentration: 10^5 ppm

for *C. intermedia*

$$Q_{O_2}(DH)_{max} = 5 \times 10^{-3} \text{ (mole } O_2) \cdot (\text{g cell})^{-1} \cdot h^{-1}$$

and for *C. tropicalis*

$$Q_{O_2}(DH)_{max} = 5.3 \times 10^{-3} \text{ (mole } O_2) \cdot (\text{g cell})^{-1} \cdot h^{-1}.$$

$Q_{O_2}(DH)_{max}$ was not equal to $Q_{O_2}(AH)_{max}$ for *C. tropicalis.* This result suggests that accommodation-form n-tetradecane is not formed during the measurement of the oxygen uptake rate on drop-form n-tetradecane.

The rates of oxygen uptake by n-tetradecane-grown cells were also measured. The results are shown in Fig. 11 for *C. intermedia* and in Fig. 12 for *C. tropicalis.* The following values for $Q_{O_2}(DH)_{max}$ were obtained for n-tetradecane-grown cells:
for *C. intermedia*

$$Q_{O_2}(DH)_{max} = 4 \times 10^{-3} \text{ (mole } O_2) \cdot (\text{g cell})^{-1} \cdot h^{-1}$$

and for *C. tropicalis*

$$Q_{O_2}(DH)_{max} = 4 \times 10^{-3} \text{ (mole } O_2) \cdot (\text{g cell})^{-1} \cdot h^{-1}$$

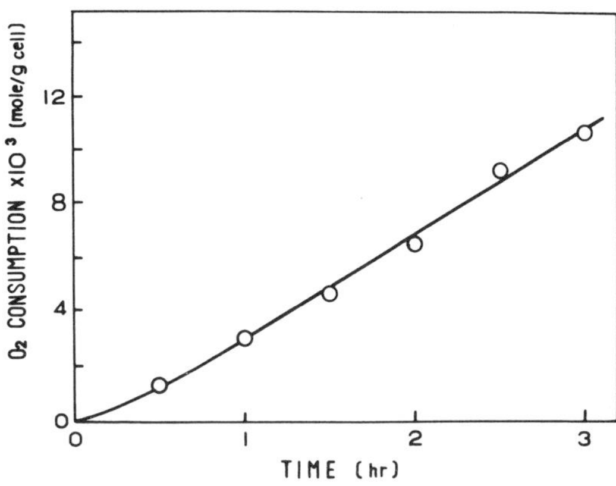

Fig. 11. Oxygen consumption by n-tetradecane-grown *C. intermedia* on drop-form n-tetradecane [8]
cell concentration: 0.02 g/l
n-tetradecane concentration: 10^5 ppm

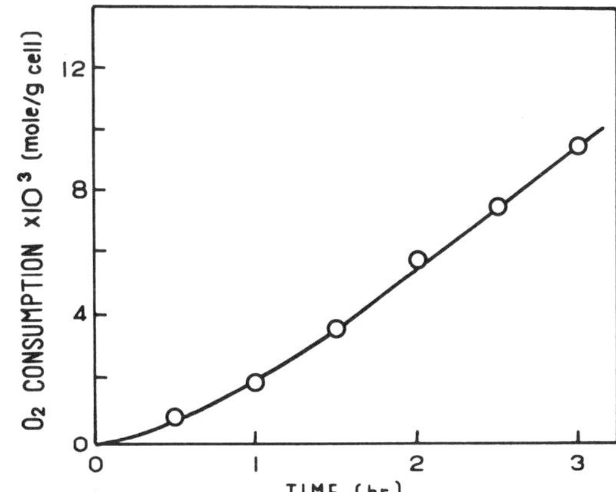

Fig. 12. Oxygen consumption by n-tetradecane-grown *C. tropicalis* on drop-form n-tetradecane [8]
cell concentration: 0.02 g/l
n-tetradecane concentration: 10^5 ppm

The rate of oxygen uptake by hydrocarbon-precultured cells on drop-form hydrocarbon was approximately equal to that of alanine-precultured cells.

For *C. intermedia,* which had a strong adhesive force with hydrocarbon, the maximum rate of oxygen uptake on accommodation-form hydrocarbon was nearly equal to that on drop-form hydrocarbon:

$$Q_{O_2}(AH)_{max} \cong Q_{O_2}(DH)_{max}$$

For *C. tropicalis,* which had a low adhesive force to hydrocarbon, the maximum rate of oxygen uptake on accommodation-form hydrocarbon was higher than that on drop-form hydrocarbon:

$$Q_{O_2}(AH)_{max} > Q_{O_2}(DH)_{max}.$$

It is considered, from the above results, that microorganisms with high affinity for hydrocarbon can utilize the drop-form hydrocarbon almost equally as well as the accommodation-form, while microorganisms with low affinity utilize the accommodation-form more effectively.

4. Hydrocarbon Pool

Lebeault *et al.* [35] reported a dehydrogenation of n-decane in mitochondrial extracts, however, Van der Linden and Huybregtse [36] as well as Liu and Johnson [37] have found alkane oxidizing enzymes localized in the cytoplasmic membrane. Ludvik *et al.* [38] observed the following ultrastructural features of a strain of *Candida lipolytica* on electron micrographs:

1) The surface of the yeast cell wall after growth on hydrocarbons is covered with a thin layer of hydrocarbons which penetrate through the cell wall to the cell membrane. The accumulation of hydrocarbons is especially marked in yeast cells grown on gas oil. Hydrocarbons accumulate on the surface of the cytoplasmic membrane.

2) The cytoplasmic membrane of cells grown on hydrocarbons is always thicker and clearly visible and contains deep invaginations and digital projections which represent an increase of the surface of the cytoplasmic membrane. Pinocytotic vesicles were frequently observed at the ends of deep invaginations, suggesting the possibility of an active translocation of hydrocarbons into the cytoplasm.

3) Yeast cells grown on hydrocarbons contain more abundant endoplasmic reticulum.

4) Cells grown on media with hydrocarbons contain more fat vacuoles than do cells grown on a glucose-containing medium.

5) Yeast cells grown on hydrocarbons have more mitochondria which frequently contain an intramitochondrial vacuole.

6) The cell wall of these yeasts is thinner than for cells grown on glucose.

7) The cytoplasm of cells grown on hydrocarbons is more electron-dense and contains more ribosomes.

8) Cells on glucose contain numerous glycogen granules, whereas the hydrocarbon grown cells contain less polysaccharide and more fat vacuoles.

From these observations, the authors conclude that hydrocarbons penetrate the cell wall of *C. lipolytica* and are concentrated at the surface of the cytoplasmic membrane bringing about numerous morphological changes of the cell and that, furthermore, the cytoplasmic membrane seems to play an important role in the metabolism of hydrocarbons as well as in their transport into the cell. In addition, Munk, Dostálek and Volfová [39] demonstrated, on electron micrographs, the penetration of hydrocarbons into the yeast cell and estimated the velocity as well as reversibility of this process by using tritium-traced hexadecane.

Kennedy and Finnerty [40] have observed an intracytoplasmic accumulation of hydrocarbon in *Micrococcus cerificans,* indicating that the alkane-oxidizing enzyme was located within the cell. Volfová *et al.* [41] showed that protoplasts of *C. lipolytica* were unable to assimilate hydrocarbons. However, Lebeault *et al.* [42] demonstrated that protoplasts of *C. tropicalis* oxidized decane and tetradecane. It appears that the location of the enzyme responsible for hydrocarbon assimilation is different depending on the kind of microorganism involved.

The author *et al.* [45] has estimated the hydrocarbon pool in and on the cell for *C. intermedia* IFO 0761 with high affinity for hydrocarbon and *C. tropicalis* ATCC 20336 with low affinity for hydrocarbon, measuring the oxygen uptake rate under the starvation of a carbon source. The results are shown in Figs. 13 and 14. The oxygen uptake rate of *C. intermedia* has a maximum value, 4.8×10^{-3} mole · (g cell)$^{-1}$ · h^{-1}, in the early period of starvation and then decreased gradually as shown in Fig. 13. That maximum value was approximately equal to the maximum oxygen uptake rate for *C. intermedia* shown in Section 3. The oxygen uptake rate of *C. tropicalis* did not show the same maximum value as that which was shown in Section 3. It is inferred that *C. intermedia* contains a pool of hydrocarbon corresponding to the maximum oxygen uptake rate in the early period after removal of the cells from the hydrocarbon medium, while *C. tropicalis* contains a smaller hydrocarbon pool than that corresponding to its maximum oxygen uptake rate.

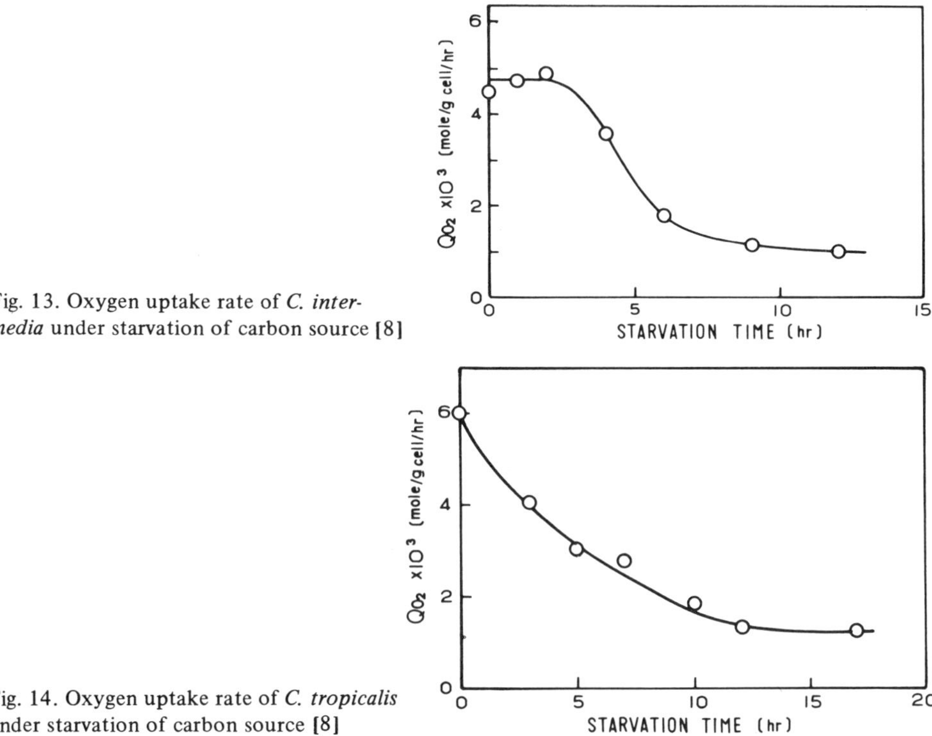

Fig. 13. Oxygen uptake rate of *C. intermedia* under starvation of carbon source [8]

Fig. 14. Oxygen uptake rate of *C. tropicalis* under starvation of carbon source [8]

5. Growth Kinetics of Microorganisms with Low Affinity for Liquid Hydrocarbon

The microorganisms with low affinity for liquid hydrocarbon utilize the accommodation-form hydrocarbon more effectively than the drop-form hydrocarbon, as shown in Section 3. Aiba *et al.* [9] proposed the following Monod-type model for the microbial growth on liquid hydrocarbon, assuming that the most susceptible oil drops to the microbial uptake are in accommodation form so that

$$\mu = \mu_{max} \frac{S^*}{K_S + S^*} \tag{1}$$

where S* designates the concentration of accommodated oil. Assuming that it is proportional to the surface area of the oil drops, S* is then correlated with operating conditions as follows:

$$S^* = \beta N^{1.2} D_i^{0.8} \left(\frac{H}{T}\right)^{-1.2} \phi^{0.24} \frac{\rho}{\sigma}^{0.6}. \tag{2}$$

If the value of S* is in a fairly small range compared with K_S, the specific growth rate, as expressed using Eqs. (1) and (2), is then:

$$\mu = \frac{\mu_{max}}{K_S} \beta N^{1.2} D^{0.8} \left(\frac{H}{T}\right)^{-1.2} \phi^{0.24} \left(\frac{\rho}{\sigma}\right)^{0.6}, \tag{3}$$

where β = proportionality constant.
Based on the assumption that cell growth is governed by the extent of probable cell attachment to the surface of the oil-droplets, Moo-Young and Shimizu [25] proposed the following Monod-type model:

$$\mu = \mu_{max} \frac{P}{K_P + P}, \tag{4}$$

where P is a measure of the potential for the accommodation of cells on the oil-droplet surface, i.e. the number of cells per unit volume of dispersion which can be attached to droplets, and K_P is the saturation constant for the potential. P is expressed by the following equation:

$$P = \frac{\overline{A}_P}{\overline{a}} = \frac{2\sqrt{3}}{3} \pi n_P (2.5 + \overline{d}^*) d^*. \tag{5}$$

When the diameter of the oil droplets is much smaller than that of the cells, the specific growth rate can be derived from Eqs. (4) and (5) as

$$\mu = \mu_{max} \frac{\dfrac{S}{\overline{d}_0^{*2}}}{\beta_1 + \dfrac{S}{\overline{d}_0^{*2}}}, \tag{6}$$

where

$$\beta_1 = \frac{\alpha^2 K_P \rho_P \overline{d}_C^3}{10\sqrt{3}},$$

and it is assumed that the oil droplet size in the culture system is proportional to the droplet size in the equivalent nonculture system, as suggested by Calderbank [43], that is

$$d_P = \alpha \, d_{PO}, \tag{7}$$

where α = proportionality constant.
Assuming that a correlation of Vermeulen *et al.* [44], i.e. Eq. (8), is available for liquid-liquid dispersions in agitation systems, the growth rate is then correlated with the operating conditions in the same way as Eq. (9) (Moo-Young and Shimizu [25])

$$\frac{\bar{d}_{PO}}{D_i} = \frac{Cf_\phi}{N_{We}^{0.6}}, \tag{8}$$

and

$$\mu = \mu_{max} \frac{SN^{2.4} \, D_i^{1.6}}{\beta_2 + SN^{2.4} \, D_i^{1.6}}, \tag{9}$$

where C = an empirical constant, f_ϕ = constant, depending on dispersed oil-phase hold up and

$$\beta_2 = \frac{\beta_1 (Cf_\phi)^2 \sigma^{1.2}}{\bar{d}_C^2 \, \rho_{aq}^{1.2}}.$$

Furthermore, Moo-Young and Shimizu [25] obtained a relationship between oil droplet size and power consumption for agitation systems, Eq. (10), and correlated the specific growth rate with the power consumption per unit volume, Eq. (11), as follows:

$$\bar{d}_{PO} = \frac{0.023}{\left(\frac{P_g}{V}\right)^{0.4}} \tag{10}$$

and

$$\mu = \mu_{max} \frac{S\left(\frac{P_g}{V}\right)^{0.8}}{\beta_3 + S\left(\frac{P_g}{V}\right)^{0.8}} \tag{11}$$

where

$$\beta_3 = \frac{(0.023)^2 \beta_2}{\bar{d}_C^2}.$$

Equations (9) and (11) show good agreement for batch and continuous fermentations with the experimental data of Moo-Young *et al.* using *Candida lipolytica* ATCC 8661 [24, 25] and Aiba *et al.* using *Candida guilliermondii* Y-7 and Y-8 [9].

Yoshida and Yamane [28] showed that a Monod-type equation could correlate the specific growth rate in the continuous fermentor with the concentration of accommodation-form hydrocarbon. They also claimed that the same equation could also be fitted to the data for the conventional-type batch culture in the fermentor in which an oil phase as well as an aqueous phase exist, provided that the hydrocarbon concentration in the aqueous phase, excluding oil drops, is employed as the substrate concentration.

Based on the assumption that the growth occurs on the soluble alkanes and submicron alkane droplets and that the metabolite produced by the growing cells helps the dissolution of liquid alkanes in the aqueous medium, Chakravarty *et al.* [30] presented a kinetic model for the microbial growth on liquid n-alkanes. The model fits well with growth data for batch and continuous cultures reported by Moo-Young *et al.* [24, 25] as well as Blanch and Einsele [2]. The model also explains the differences between the relative length of the exponential and the linear phases of the growth.

According to the above models, the growth rate is considerably influenced by operating conditions for microorganisms with low affinity for liquid hydrocarbon, which primarily utilize the accommodation-form hydrocarbon.

Wang and Ochoa [19] also showed that the specific growth rate is directly related to the specific hydrocarbon interfacial area, which in turn is directly related to the impeller speed, the hydrocarbon concentration and the surfactant concentration. Blanch and Einsele [2] investigated the kinetics of growth by *Candida tropicalis* on pure n-hexadecane and showed that, while exponential growth was independent of stirrer speed, linear growth was indeed determined by the stirrer speed.

6. Growth Kinetics of Microorganisms with High Affinity for Liquid Hydrocarbon

The Microorganisms with high affinity for liquid hydrocarbon can utilize the drop-form hydrocarbon as well as the accommodation-form as was shown in Section 3. In this section, the growth kinetics of microorganisms with high affinity for liquid hydrocarbon is discussed, using the experimental results of the author *et al.* [45] for *Candida intermedia* IFO 0761 as an example.

6.1 Transfer of Substrate During Clump Formation Accompanied Growth

Microorganisms with high affinity for liquid hydrocarbon often form clumps with hydrocarbon drops during growth (Mimura *et al.* [1]; Blanch and Einsele [2]; Einsele *et al.* [31] and Nakahara *et al.* [33]). *C. intermedia* IFO 0761 adheres tightly to the surface of the hydrocarbon drop and the microbes on the hydrocarbon drops form clumps during growth, as shown in Section 2. The transfer of substrates other than hydrocarbon was investigated in the growth which accompanied clump formation. This was done so as to assess which substrate-transfer step limits the growth rate, knowledge of which is important for the formulation of a kinetic model of growth. The uptake rate of ^{32}P-H_3PO_4 from a medium containing n-tetradecane and glucose was compared with that of a glucose medium without n-tetradecane. The maximum uptake rate of ^{32}P-H_3PO_4 from the glucose and n-tetradecane media was 250 and 90 cpm · (mg cell)$^{-1}$ · min^{-1}, respectively. The maximum specific growth rate on the glucose medium was about three times higher than that on the n-tetradecane medium, as shown in Table 3.

Table 3. Maximum uptake rate of $^{32}P\text{-}H_3PO_4$ and maximum specific growth rate for *C. intermedia* at 30 °C [8]

Carbon source	n-tetradecane	glucose
Maximum uptake rate of $^{32}P\text{-}H_3PO_4$ (cpm/mg cell/min)	90	250
Maximum specific growth rate (h^{-1})	0.11	0.31

The uptake rate of $^{32}P\text{-}H_3PO_4$ shifts upward with the addition of glucose to the n-tetradecane medium, as shown in Fig. 15. Figure 15 indicates that the uptake rate of $^{32}P\text{-}H_3PO_4$ was shifted upward by 280 cpm · (mg cell)$^{-1}$ · min^{-1} during the exponential growth phase. During the decreasing growth phase, the uptake rate of $^{32}P\text{-}H_3PO_4$ was shifted up by 250 cpm · (mg cell)$^{-1}$ · min^{-1} by the addition of glucose to the n-tetradecane medium, as shown in Fig. 16. The uptake rate of $^{32}P\text{-}H_3PO_4$, after the addition

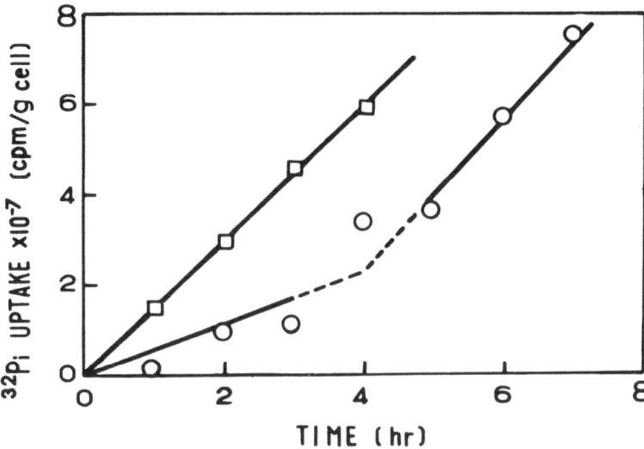

Fig. 15. $^{32}P\text{-}H_3PO_4$ uptake in exponential growth phase [8]
- o n-tetradecane medium with the addition of glucose after 4 h
- □ glucose medium

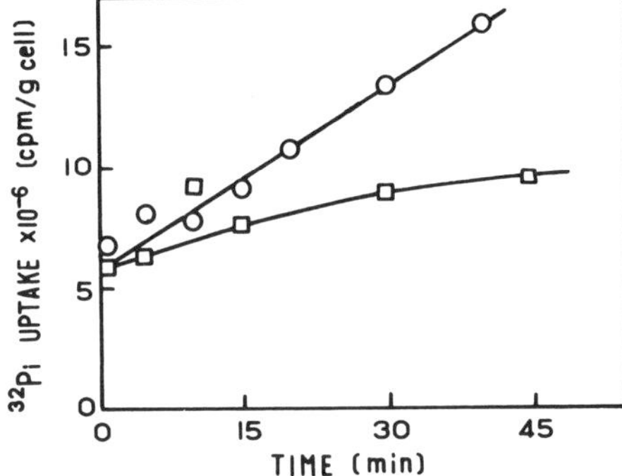

Fig. 16. $^{32}P\text{-}H_3PO_4$ uptake in decreasing growth phase [8]
- o n-tetradecane medium with addition of glucose at time zero
- □ n-tetradecane medium

of glucose to the n-tetradecane medium, was approximately equal to that on the glucose medium. It is inferred that the transfer of $^{32}P\text{-}H_3PO_4$ and other water-soluble substrates to the cell are not the rate-limiting step in the growth of *C. intermedia* which is accompanied by clump formation with n-tetradecane. Therefore, the transfer of n-tetradecane is considered to be the limiting step for the growth rate and this is controlled by the concentration of n-tetradecane.

6.2 The Effect of Operating Conditions on the Growth Rate

The effect of agitation speed on the oxygen supply rate was investigated, using a 15 l jar fermentor, in order to obtain minimum necessary agitation speed of the oxygen supply needed for growth. The oxygen supply rate increased as the result of the agitation speed, as shown in Fig. 17. The volumetric oxygen transfer coefficient, $k_L a$, decreased due to cell growth as well as with the addition of an antifoamer, silicone KM 70 (Shin-etsu Chemical Industries, Ltd., Japan). The effect of the antifoamer on the volumetric oxygen transfer coefficient is shown in Fig. 18. This figure indicates that the oxygen transfer rate is decreased by the antifoamer for concentrations up to 10 ppm but remains approximately constant for higher concentrations. The volumetric oxygen transfer coefficient in the culture filtrate with antifoamer was approximately equal to the value for the aqueous solution of the antifoamer, as shown in Fig. 17. Therefore, the culture used included the antifoamer.

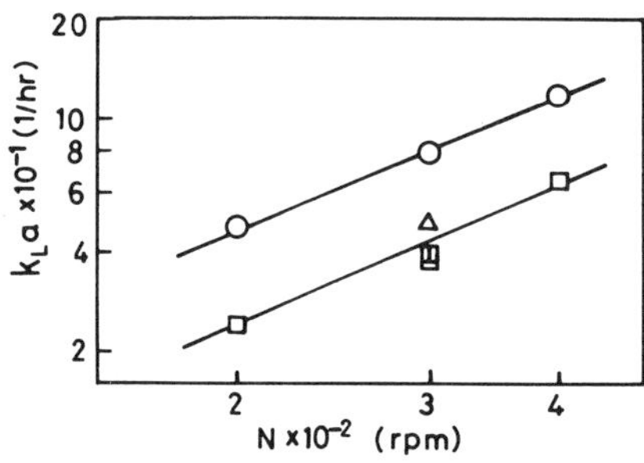

Fig. 17. Oxygen transfer rate in liquid phase of 15-l jar fermenter at aeration rate of 1 vvm and 30 °C [8]

○ in water without antifoamer

□ in water with 50 ppm of antifoamer

△ in culture filtrate

⊡ in culture filtrate with 50 ppm of antifoamer

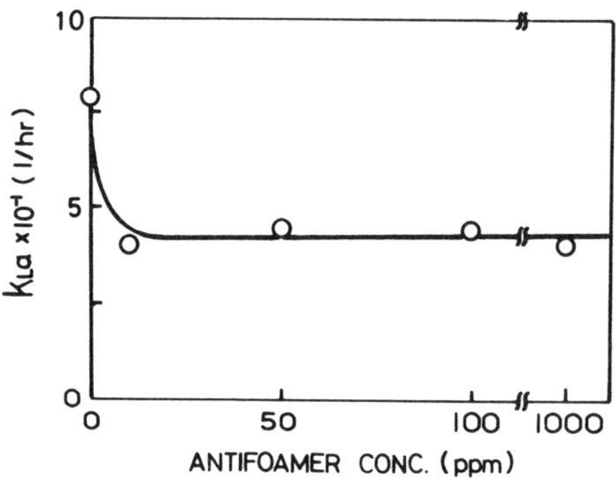

Fig. 18. Effect of antifoamer on oxygen transfer in liquid phase of 15-l jar fermenter at aeration rate of 1 vvm and agitation speed of 300 rpm and 30 °C [8]

When an initial concentration 2000 ppm of n-tetradecane was used, the maximum cell concentration was $2 \cdot g\,l^{-1}$ and the maximum oxygen uptake rate was $4.8 \times 10^{-3} \cdot$ (mole O_2) $\cdot$ (g cell)$^{-1} \cdot h^{-1}$. From the results of Fig. 17, the minimum agitation speed, necessary for the oxygen supply to provide for maximum oxygen consumption, was estimated to be about 300 rpm when the air was supplied at 1 atmosphere and 1 vvm and about 170 rpm when the air was supplied at 2 atmospheres and 1 vvm, assuming that the solubility of oxygen in the basal medium is equal to that in water. The cultures were grown with the oxygen supply rate higher than the maximum rate of oxygen uptake by the cells; the agitation speeds used were 200, 300, and 400 rpm ant the air was supplied at 1 atmosphere and 1 vvm at the agitation speeds of 300 and 400 rpm and at 2 atmospheres and 2 vvm at the agitation speed of 200 rpm. The initial concentration of n-tetradecane was 2000 ppm.

The results of the experiments are shown in Figs. 19–21. The rates of acid production were proportional to the rates of cell growth, though a time lag in acid production was observed. Consequently, the growth rate could be estimated from the rate of acid production. Figure 22 shows the effects of operating conditions on the specific growth rate. The maximum specific growth rate was approximately the same under the three sets of operating conditions given in Figs. 19–21: 0.17 h^{-1}, 0.11 h^{-1}, and 0,13 h^{-1}, respectively, as shown in Fig. 22. From the above results it is concluded that these operating conditions, with respect to the oxygen supply, are adequate for the growth of *C. intermedia* on n-tetradecane. Therefore, for microorganisms with high affinity for liquid hydrocarbon, the growth rate is considered not to be influenced by the operating conditions, as long as the oxygen supply is sufficient for growth.

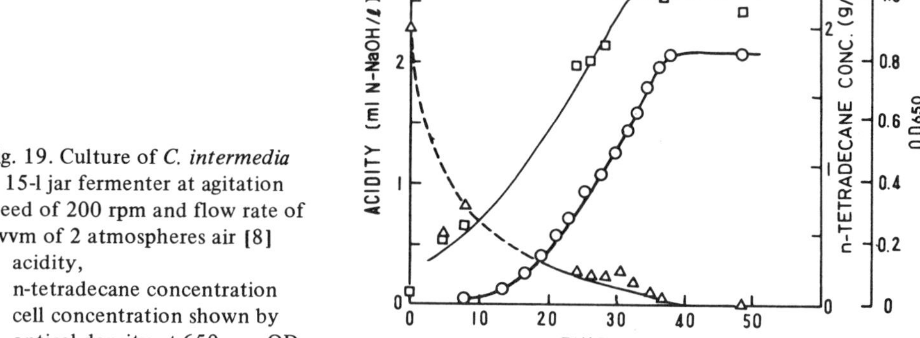

Fig. 19. Culture of *C. intermedia* in 15-l jar fermenter at agitation speed of 200 rpm and flow rate of 2 vvm of 2 atmospheres air [8]
○ acidity,
△ n-tetradecane concentration
□ cell concentration shown by optical density at 650 nm, OD_{650}

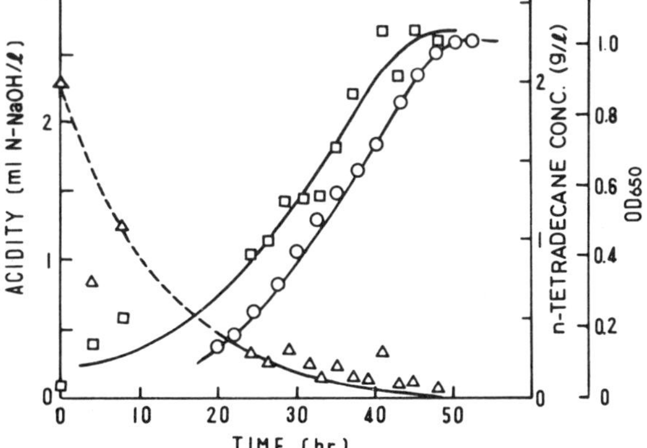

Fig. 20. Culture of *C. intermedia* in 15-l jar fermenter at agitation speed of 300 rpm and flow rate of 1 vvm of 1 atmosphere [8]
○ acidity
△ n-tetradecane concentration
□ cell concentration shown by optical density at 650 nm, OD_{650}

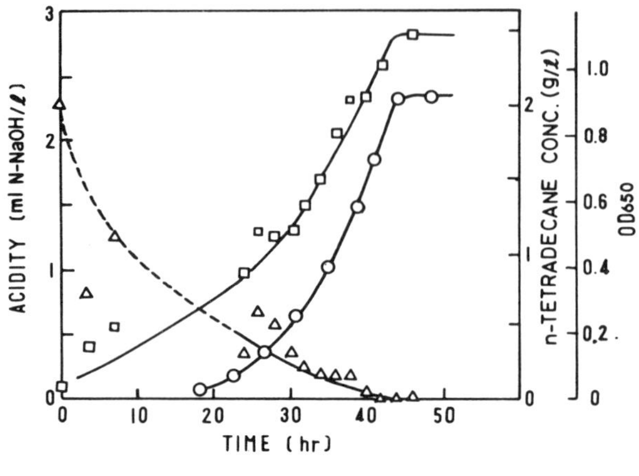

Fig. 21. Culture of *C. intermedia* in 15-l jar fermenter at agitation speed of 400 rpm and flow rate of 1 vvm of 1 atmosphere air [8]
○ acidity
△ n-tetradecane concentration
□ cell concentration shown by optical density at 650 nm, OD_{650}

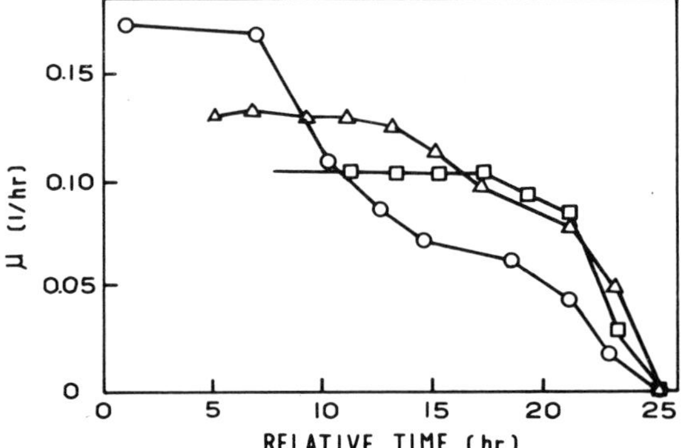

Fig. 22. Specific growth rates under three sets of operating conditions [8]

	○	□	△
agitation speed (rpm)	200	300	400
aeration rate (vvm)	2	1	1
air pressure (atm.)	2	1	1

6.3 Kinetic Model for Growth

Lineweaver-Bulk plots of specific growth rates and n-tetradecane concentrations for the three sets of operating conditions gave an almost straight line for *C. intermedia* IFO 0761, as shown in Fig. 23. The maximum specific growth rate obtained from the Lineweaver-Bulk plot in Fig. 23 was 0.4 h^{-1}, but this was quite different from the experimental value shown in Fig. 22. Therefore, Monod's model is not valid for the growth rate of *C. intermedia* on a n-tetradecane medium.

C. intermedia has a strong affinity for hydrocarbon and is considered to contain a hydrocarbon pool in and/or on the cell, as inferred from Fig. 13. It is, therefore, concluded that the growth rate of *C. intermedia* on hydrocarbon is a function of the hydrocarbon pool. The hydrocarbon pool is related to the specific concentration of hydrocarbon per unit cell mass in the medium by the term $S \cdot X^{-1}$. Thus the relationship between the specific growth rate and the specific concentration of hydrocarbon per unit cell mass, $S \cdot X^{-1}$, was investigated. The results are shown in Fig. 24, which indicates that the saturation kinetics is valid for the specific growth rate and the specific concentration of hydrocarbon per unit cell mass. The Lineweaver-Bulk plots of specific

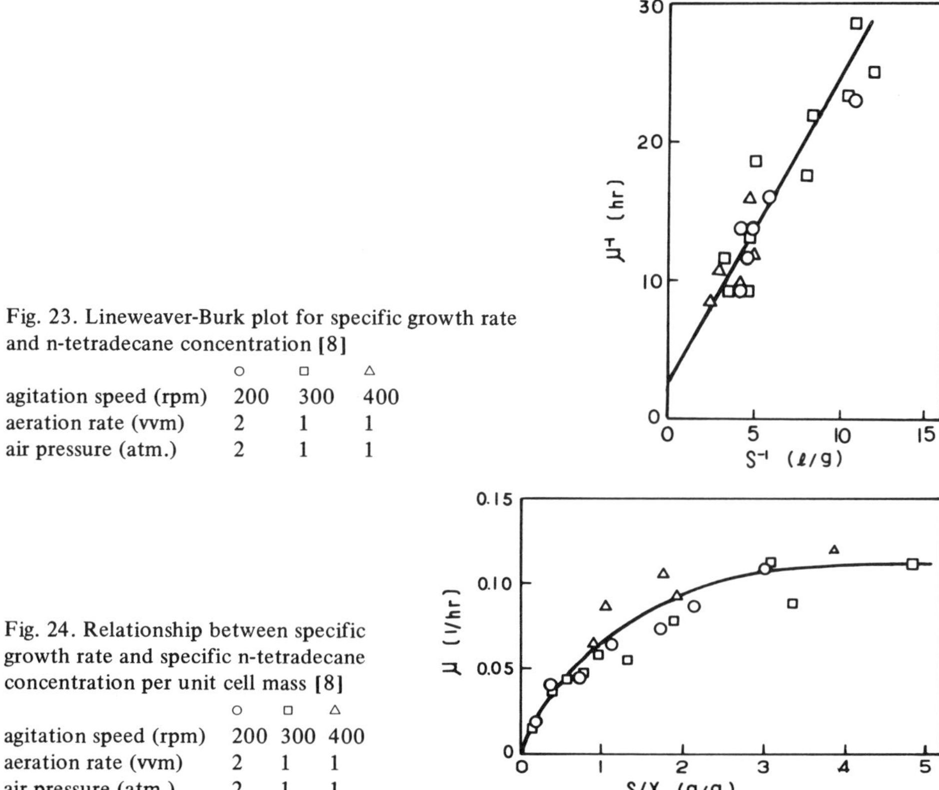

Fig. 23. Lineweaver-Burk plot for specific growth rate and n-tetradecane concentration [8]

	○	□	△
agitation speed (rpm)	200	300	400
aeration rate (vvm)	2	1	1
air pressure (atm.)	2	1	1

Fig. 24. Relationship between specific growth rate and specific n-tetradecane concentration per unit cell mass [8]

	○	□	△
agitation speed (rpm)	200	300	400
aeration rate (vvm)	2	1	1
air pressure (atm.)	2	1	1

growth rate and specific hydrocarbon concentration for the three sets of operating conditions all give almost a straight line, as shown in Fig. 25. The maximum specific growth rate obtained from the results in Fig. 25 was 0.12 h^{-1}, this agreed approximately with the experimental value shown in Fig. 22. It is, therefore, concluded that the specific

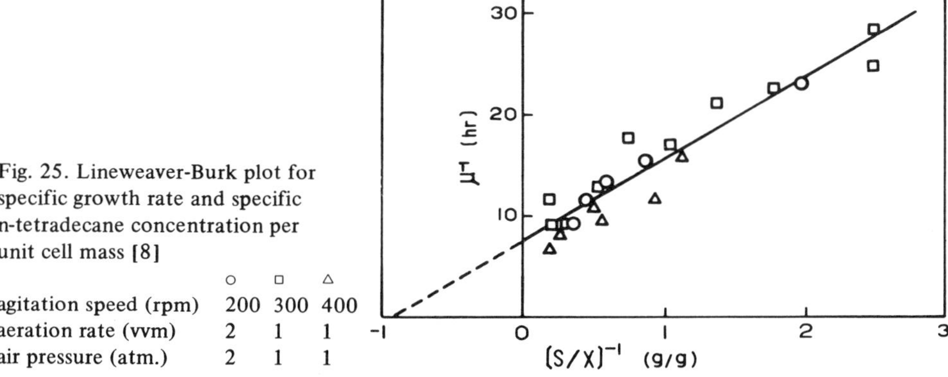

Fig. 25. Lineweaver-Burk plot for specific growth rate and specific n-tetradecane concentration per unit cell mass [8]

	○	□	△
agitation speed (rpm)	200	300	400
aeration rate (vvm)	2	1	1
air pressure (atm.)	2	1	1

growth rate of *C. intermedia* on hydrocarbon is expressed by the following equation:

$$\mu = \frac{1}{X}\frac{dX}{dt} = \frac{\mu_{max}\frac{S}{X}}{K_S + \frac{S}{X}}. \qquad (12)$$

The specific rate of oxygen uptake is related to the specific hydrocarbon concentration, $S \cdot X^{-1}$, as shown in Figs. 26 and 27. These results also indicate that the above-mentioned saturation kinetics are valid for the specific growth rate of *C. intermedia* and the specific hydrocarbon concentration, $S \cdot X^{-1}$, as the specific rate of oxygen uptake was proportional to the specific growth rate, as shown in Fig. 4.

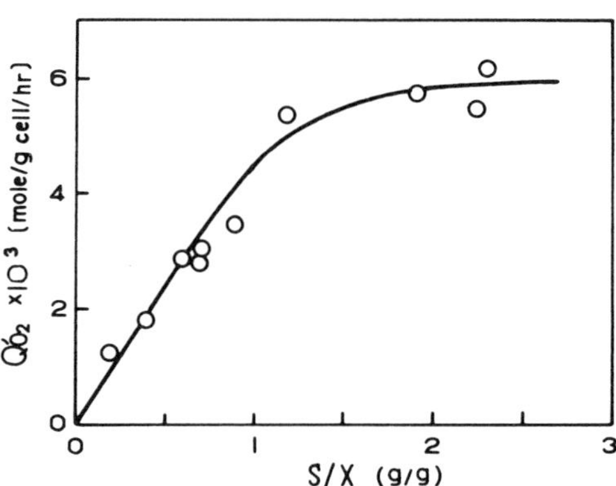

Fig. 26. Relationship between oxygen uptake rate and specific n-tetradecane concentration per unit cell mass [8]

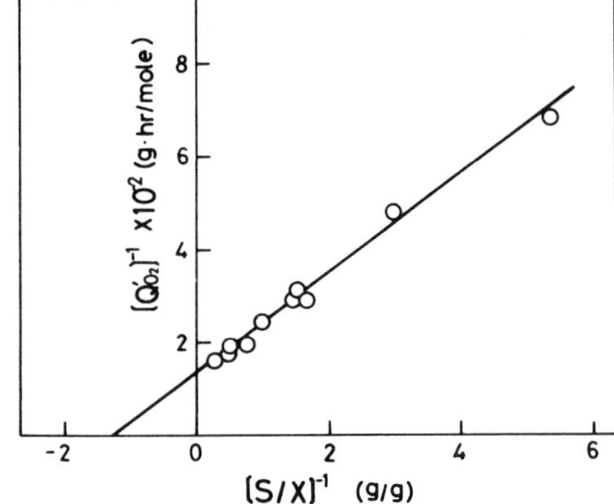

Fig. 27. Lineweaver-Burk plot for oxygen uptake rate and specific n-tetradecane concentration per unit cell mass [8]

7. Concluding Remarks

The mechanism of liquid hydrocarbon uptake by microorganisms depends upon the properties of the microorganisms and the kinds of hydrocarbons. The microorganisms with high affinity for hydrocarbon can utilize the large drop-form hydrocarbon as well as the submicron accommodation-form, while the microorganisms with low affinity for hydrocarbon utilize the accommodation-form hydrocarbon more effectively than the drop-form. The quantity of dissolved hydrocarbon utilized is negligible compared with the quantity of drop-form and accommodation-form hydrocarbons utilized, when comparatively longer chain hydrocarbons, such as decane, tetradecane and hexadecane are used as a substrate. The relative contributions of drop-form hydrocarbon, accommodation-form hydrocarbon and dissolved hydrocarbon to the microbial utilization depend upon the properties of the microorganisms, the kinds of hydrocarbons and the experimental conditions. In batch fermentation, those relative contributions may change as the fermentation proceeds. It was reported by several researchers (Mimura *et al.* [1]; Tanaka, Fukui [46]; Wang and Ochoa [19]; Whitworth *et al.* [47] that the surfactants have an important role with respect to the microbial growth on liquid hydrocarbon.

The growth rate of the microorganism with low affinity for hydrocarbon is considerably influenced by the operating conditions. The growth rate of the microorganism with high affinity is scarcely influenced by the operating conditions provided the oxygen supply is sufficient to sustain growth. The saturation kinetics is valid for the specific growth rate and the specific hydrocarbon concentration for microorganism with high affinity for hydrocarbon.

A mannan-fatty acid-complex is considered to increase the lipophility of the cell surface and the affinity of the cell to hydrocarbon for *Candida tropicalis* ATCC 32113. There are many reports of increased lipid content of both bacteria and yeast on hydrocarbon (Johnson [14]; Mizuno *et al.* [48]; Dunlap and Perry [49]; Nyns *et al.* [50]; Koronelli [51]; Hug *et al.* [5]. The role of cellular lipids in hydrocarbon assimilation has been discussed by several researchers. Dunlap and Perry [49] proposed that on the basis of the solubility of hydrocarbons the cellular lipids play an important role in hydrocarbon assimilation; intermediates of alkane degradation provided a solvent for the insoluble alkanes. Vestal and Perry [52] also suggested that an increased lipid content was necessary for the uptake and the accumulation of lipophilic substrates.

The functional role of cellular lipids in hydrocarbon assimilation was discussed by Hug *et al.* [5]. When grown on hydrocarbons, the yeast *Candida tropicalis* contained twice as much lipid as when grown on glucose. In the transient continuous culture phase, following a substrate change form glucose to hexadecane, an adaptation occured. The lipid concentration per cell increased greatly during that transient phase. The cause of that adaptation phase was assumed to be due to both induction of the enzymes required for hexadecane oxidation and the necessity of transporting this substrate to the site of enzymatic action. These authors [5] proposed that the role of lipid in the hydrocarbon assimilation process is to provide a hydrophobic region through which the lipophilic substrate may be transported, i.e. to act as a solvent for the hydrocarbon.

The carbon energy reserve metabolism for *C. tropicalis* growing on glucose and on hydrocarbons was investigated by Kaeppeli *et al.* [53]. They showed that glycogen was markedly accumulated in *C. tropicalis* growing on glucose, that the same effect was caused by a N-free medium and that the lipid content did not show any significant change in either case. On the hydrocarbon substrate, lipid increased as substrate availability decreased whereas glycogen accumulation was only slight. However, the increase of lipid content on hydrocarbons did not reach the same level of accumulation as glycogen on glucose. In an N-free medium, both glycogen and lipids were accumulated. From these results the authors suggested that glycogen is not substituted by lipids as the carbon energy reserve on a hydrocarbon substrate.

The relationships between the function and structure of the n-alkane-utilizing yeast cells were investigated by Hirai *et al.* [54], Osumi *et al.* [55, 56] and Teranishi *et al.* [57, 58]. They observed many interesting features of the physiological activity and the ultrastructure of the cells: morphological change depending upon the chain length of n-alkane substrate, development of microbodies relevant to a marked increase of catalase activity and so on.

There is other interesting research going on with respect to physiology and morphology of microorganisms on hydrocarbon. There are still many unsolved problems and they have to be further investigated. The growth kinetics of microorganisms on hydrocarbon will have to be further discussed on the basis of the results of those investigations.

Nomenclature

$\overline{A}_P$ mean interfacial area between dispersed ans continuous phase per unit volume of dispersion, cm^{-1}
$\overline{a}$ mean surface area occupied per cell on oil droplet, cm^2
D_i agitator diameter, cm
$\overline{d}_C$ Sauter mean diameter of cells, cm
$\overline{d}_P$ Sauter mean diameter of oil droplets, cm
$\overline{d}_{PO}$ Sauter mean diameter of oil droplets in oil-basal salt solution, cm
$\overline{d}^*$ $\overline{d}_P \cdot \overline{d}_C^{1}$
d_0^* $\overline{d}_{PO} \cdot \overline{d}_C^{1}$
H liquid depth in fermentor vessel, cm
K_S saturation constant for growth kinetic model, $g \cdot l^{-1}$ in Eq. (1), $g \cdot g^{-1}$ in Eq. (12)
k_La volumetric oxygen transfer coefficient for liquid phase, h^{-1}
N agitation speed, rps or rpm
N_{We} Weber number = $N^2 \cdot D_i^3 \cdot \rho_{aq} \cdot \sigma^{-1}$
n_P number of oil droplets per unit volume (cm^{-3}) of dispersion,
P_g agitator power consumption in gassed fermentor, hp
Q_{O_2} oxygen uptake rate, (mole O_2) $\cdot$ (g cell)$^{-1} \cdot h^{-1}$
Q'_{O_2} (oxygen uptake rate) – (endogenous respiration rate), (mole O_2) $\cdot$ (g cell)$^{-1} \cdot h^{-1}$
S concentration of oil in medium, $g \cdot l^{-1}$
S* concentration of accommodated oil in medium, $g \cdot l^{-1}$
T diameter of reactor, cm
t growth time, h
V working volume of reactor, m^3
X cell concentration, $g \cdot l^{-1}$
μ specific growth rate, h^{-1}
μ_{max} maximum specific growth rate, h^{-1}
ρ density of continuous phase, $g \cdot cm^{-3}$
ρ_{aq} density of aqueous phase, $g \cdot cm^{-3}$
ρ_0 density of oil phase, $g \cdot cm^{-3}$
σ interfacial tension, dyne $\cdot$ cm^{-1}
ϕ volume fraction of oil in medium

References

1. Mimura, A., Watanabe, S., Takeda, I.: J. Ferment. Technol. **49**, 255 (1971).
2. Blanch, H. W., Einsele, A.: Biotechnol. Bioeng. **15**, 861 (1973).
3. Einsele, A., Schneider, H., Fiechter, A.: J. Ferment. Technol. **53**, 241 (1975).
4. Kaeppeli, O., Fiechter, A.: Biotechnol. Bioeng. **18**, 967 (1976).
5. Hug, H., Blanch, H. W., Fiechter, A.: Biotechnol. Bioeng. **16**, 965 (1974).
6. Kaeppeli, O.: Dissertation, Eidgenössische Technische Hochschule Zürich, 1976.

7. Osumi, M., Fukuzumi, F., Yamada, N., Nagatani, T., Teranishi, Y., Tanaka, A., Fukui, S.: J. Ferment. Technol. **53**, 244 (1975).
8. Miura, Y., Okazaki, M., Hamada, S., Murakawa, S., Yugen, R.: Biotechnol. Bioeng. **19**, 701 (1977).
9. Aiba, S., Haung, K. L., Moritz, V., Someya, J.: J. Ferment. Technol. **47**, 211 (1969).
10. Erdstieck, B., Rietema, K.: Antonie van Leeuwenhoek, **35**, Supplement, Yeast Symp., F19 (1969).
11. Yoshida, F., Yamane, T., Yagi, H.: Biotechnol. Bioeng. **13**, 215 (1971).
12. Yoshida, F., Yamane, T.: Biotechnol. Bioeng. **13**, 691 (1971).
13. Chakravarty, M., Amin, P. M., Singh, H. D., Baruah, J. N., Iyengar, M. S.: Biotechnol. Bioeng. **14**, 61 (1972).
14. Johnson, M. J.: Chem. Industry 1532 (1964).
15. Aiba, S., Haung, K. L.: Chem. Eng. Japan **34**, 868 (1970).
16. Dunn, I. J.: Biotechnol. Bioeng. **10**, 891 (1968).
17. Erickson, L. E., Humphrey, A. E., Prokop, A.: Biotechnol. Bioeng. **11**, 449 (1969).
18. Prokop, A., Ludvik, M., Erickson, L.: Biotechnol. Bioeng. **14**, 587 (1972).
19. Wang, D. I. C., Ochoa, A.: Biotechnol. Bioeng. **14**, 345 (1972).
20. Katinger, H. W. D.: Advances in Microbial Engineering, Part I (Biotechnol. Bioeng. Symp., No. 4), Ed. Sikyta, B., Prokop, A., Novak, M., Eds., New York: Wiley, Interscience, p. 485, 1973.
21. Erickson, L. E., Humphrey, A. E.: Biotechnol. Bioeng. **11**, 467 (1969).
22. Erickson, L. E., Humphrey, A. E.: Biotechnol. Bioeng. **11**, 489 (1969).
23. Erickson, L. E., Fan, L. T., Shah, P. S., Chen, M. S. K.: Biotechnol. Bioeng. **12**, 713 (1970).
24. Moo-Young, M., Shimizu, T., Whitworth, D. A.: Biotechnol. Bioeng. **13**, 741 (1971).
25. Moo-Young, M., Shimizu, T.: Biotechnol. Bioeng. **13**, 761 (1971).
26. Goma, G., Pareilleux, A., Durand, G.: J. Ferment. Technol. **51**, 616 (1973).
27. Yoshida, F., Yamane, T., Nakamoto, K.: Biotechnol. Bioeng. **15**, 257 (1973).
28. Yoshida, F., Yamane, T.: Biotechnol. Bioeng. **16**, 635 (1974).
29. Hisatsuka, K., Nakahara, T., Minoda, Y., Yamada, K.: Agr. Biol. Chem. **39**, 999 (1975).
30. Chakravarty, M., Singh, H. D., Baruah, J. N.: Biotechnol. Bioeng. **17**, 399 (1975).
31. Einsele, A., Blanch, H. W., Fiechter, A.: Advances in Microbial Engineering, Part I (Biotechnol. Bioeng., Symp., No. 4), Eds. Sikyta, B., Prokop, A., Novak, M., New York: Wiley, Interscience, p. 445, 1973.
32. Bakhuis, E., Bos, P.: Antonie von Leeuwenhoek **35**, Supplement, Yeast Symp. F 47 (1969).
33. Nakahara, T., Erickson, L. E., Gutierrez, J. R.: Biotechnol. Bioeng. **19**, 9 (1977).
34. Shah, P. S., Erickson, L. E., Fan, L. T., Prokop, A.: Biotechnol. Bioeng. **14**, 533 (1972).
35. Lebeault, J. M., Roche, B., Duvnjak, Z., Azoulay, E.: Arch. Mikrobiol. **72**, 140 (1970).
36. Van der Linden, A. C., Huydregtse, R.: Antonie van Leeuwenhoek **33**, 381 (1967).
37. Liu, C. M., Johnson, M. J.: J. Bacteriol. **106**, 830 (1971).
38. Ludvik, J., Munk, V., Dostálek, M.: Experientia **24**, 1066 (1968).
39. Munk, V., Dostálek, M., Volfová, O.: Biotechnol. Bioeng. **11**, 383 (1969).
40. Kennedy, R. S., Finnerty, W. R.: the 72nd Annu. Meeting of the Amer. Soc. for Microbiol. (AMS) April 23–28, Philadelphia, Pa., 1972.
41. Volfová, O., Munk, V., Dostálek, M.: Experientia **23**, 1005 (1967).
42. Lebeault, J. M., Roche, B., Duvnjak, Z., Azoulay, E.: J. Bacteriol. **100**, 1218 (1969).
43. Calderbank, P. H.: Trans. Inst. Chem. Engrs. (London) **36**, 443 (1958).
44. Vermeulen, T., Williams, G. M., Langlois, G. E.: Chem. Eng. Progr. **51**, 85F (1955).
45. Miura, Y., Okazaki, M., Murakawa, S., Hamada, S., Ohno, K.: Biotechnol. Bioeng. **19**, 715 (1977).
46. Tanaka, A., Fukui, S.: J. Ferment. Technol. **49**, 809 (1971).
47. Whitworth, D. A., Moo-Young, M., Viswanatha, T.: Biotechnol. Bioeng. **15**, 649 (1973).
48. Mizuno, M., Shimojima, Y., Iguchi, T., Takeda, I., Senoh, S.: Agr. Biol. Chem. **30**, 606 (1966).
49. Dunlap, K. R., Perry, J. J.: J. Bacteriol. **94**, 1919 (1967).
50. Nyns, E. J., Chiang, N., Wiaux, A. C.: Antonie van Leeuwenhoek, **34**, 197 (1968).
51. Koronelli, T. V.: Mikrobiologya **37**, 984 (1968).

52. Vestal, J. R., Perry J. J.: Can. J. Microbiol. **17**, 445 (1971).
53. Kaeppeli, O., Aeschbach, H., Schneider, H., Fiechter, A.: European J. Appl. Microbiol. **1**, 199 (1975).
54. Hirai, M., Shimizu, S., Teranishi, Y., Tanaka, A., Fukui, S.: Agr. Biol. Chem. **36**, 2335 (1972).
55. Osumi, M., Miwa, N., Teranishi, Y., Tanaka, A., Fukui, S.: Arch. Mikrobiol. **99**, 181 (1974).
56. Osumi, M., Fukuzumi, F., Teranishi, Y., Tanaka, A., Fukui, S.: Arch. Microbiol. **103**, 1 (1975).
57. Teranishi, Y., Tanaka, A., Osumi, M., Fukui, S.: Agr. Biol. Chem. **38**, 1213 (1974).
58. Teranishi, Y., Kawamoto, S., Tanaka, A., Osumi, M., Fukui, S.: Agr. Biol. Chem. **38**, 1221 (1974).

Microbial Production of Hydrogen

J. E. Zajic, N. Kosaric and J. D. Brosseau
Biochemical Engineering, Faculty of Engineering Science,
The University of Western Ontario, London, Ontario, Canada

Contents

Hydrogen gas is synthesized by a rather large group of microorganisms. Of the cultures reported photosynthetic and nonphotosynthetic bacteria and algae produce large amounts under anaerobic conditions. Hydrogen gas producing cultures (*coli,* clostridial or ruminococcoid) isolated from primary sewage sludge or the rumen of cattle appear to be quite active especially when cocultured with a hydrogen utilizing microorganism. Physiological concentrations of acetyl-CoA/CoA, NADH/NAD and ferredoxin appear to be quite important in most metabolic reactions in which hydrogen is synthesized. Some of the enzymes responsible for the regulatory reactions leading to hydrogen gas appear to be phosphotransacetylase, acetate kinase, NADH: ferredoxin oxidoreductase and ferredoxin hydrogenase. In developing heterotrophic microbial processes for producing hydrogen, greater emphasis must be placed on systems which not only generate protons and electrons, but also on the capability of shunting electrons toward H_2 production rather than toward alternative reductive activities. Likewise, systems must be sought which actively carry out a reductive cleavage of water with active release of hydrogen. Photosynthetic systems need development study and more research is required for study of hydrogen production from symbionts.

Introduction

Before reviewing the literature on the bioproduction of hydrogen some of the properties of hydrogen must be presented. Hydrogen gas is an ideal fuel, not only as an alternative energy source but also as a highly efficient energy carrier [134]. The heating value of hydrogen gas per unit volume is less than other gaseous fuels ($1/3 \times CH_4$), however, the heating content per unit mass of liquid hydrogen is about 2.75 times greater than that of hydrocarbon fuels (Table 1). It has, however, great potential for use as a primary or secondary energy course, for chemical synthesis or for electrical storage and generation with fuel cells. Today, hydrogen is produced for industrial use as a chemical feedstock in the production of ammonia, methanol, refined petroleum fuels, hydrogenated vegetable and animal oils and other chemicals (see review by Gregory *et al.*, [134]). As a fuel, hydrogen has a wider range of explosive concentrations in air than does natural gas but the lower explosive limit is nearly the same for both gases. When hydrogen is burned in pure oxygen, the only end product is water, thus gaseous pollutants are not

Table 1. Comparative characteristics of synthetic fuels (after Michel, 1973)

Fuel	Heat of combustion low heating value $\left[\frac{g \cdot Cal}{g}\right]$ $\times 10^3$	$\left[\frac{kg \cdot Cal}{m^3}\right]$ $\times 10^6$	Relative fuel required to equal H_2 heat content by weight	Density of hydrogen in fuel g/l	Gas Stp g/l	Boiling point (C)	Ease of storage (relative ranking	Toxicity (relative ranking)
Hydrogen (H_2)	28.6	2.7	1.0	70.5	.080	– 253	6	1
Ammonia (NH_3)	4.4	–	6.4	124.9	.689	– 33	4	5
Hydrazine (N_2H_4)	4.0	–	7.2	142.6	–	+ 113	3	6
Methanol (CH_3OH)	4.8	–	6.0	113.7	–	+ 65	2	4
Methane (CH_4)	12.0	8.0	2.4	104.1	.657	– 162	5	2
Ethanol (C_2H_5OH)	6.4	–	4.4	104.1	–	+ 78	1	3
Gasoline (C_8H_{18})	10.6	–	2.7	112.1	–	+ 125	1	4

a problem [252]. However, this is not true when air is the source of O_2 since low concentrations of nitrous oxides are also formed.

In 1968 the amount of hydrogen produced in the United States was 65.3×10^9 cubic meters, having an energy equivalent of 741×10^{12} Btu (1.87×10^{17} gcal.). At present, hydrogen gas production costs are generally related to natural gas and petroleum costs because most of the hydrogen gas consumed in the United States is produced from gaseous or liquid hydrocarbons [335, 134]. For natural gas costing $ 1.30/2.5 $\times 10^6$ gcal., H_2 costs $ 1.45/2.5 $\times 10^6$ gcal [7]. Processes used in hydrogen gas production generally involve the reaction between hydrocarbons and steam to yield hydrogen gas (H_2), carbon monoxide (CO) and carbon dioxide (CO_2).

There are basically four processes available for the production of hydrogen gas from nonfossil primary energy sources. These include: water electrolysis, thermochemical and radiolytic processes [252]. Nuclear, either fission or fusion [347, 316], solar [472, 66] or geothermal [343] energy are potential alternative heat sources for the thermal production of hydrogen from water. Electrolytic hydrogen is presently produced competitively for industrial use, but only in areas where cheap electricity is readily available. For the economic considerations involving the use of electrolytic hydrogen as a fuel source in the future, see Gregory *et al.* (1972) [134].

A fourth process used for the production of fuels is biologically based. A primary example is the microbial formation of methane under anaerobic conditions. The history of this process is reviewed by Barker (1956) [19]; Stadtman (1967) [385]; Wolfe, (1971) [474]; Pine (1971) [312]; Taylor (1975) [415], and Zeikus (1977) [496]. Engineering theory and design are treated in Andrews and Graef (1971) [6] and Lawrence (1971) [227]. A small group of bacteria is responsible for the formation of methane. These methanogenic bacteria are readily found in anaerobic environments where organic matter is being decomposed. Such environments include swamps [474], lake sediments and rice paddies, [222], oil strata [103], and coal fields [222], digestive tract of animals, ruminants [169] and non-ruminants [67] and in sewage sludge digestion [409]. The microbial food chain that exists in these mixed culture environments is presented in Fig. 1. Cellulose or other

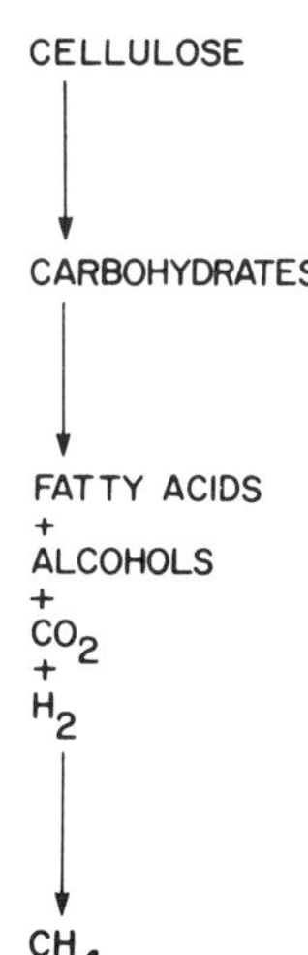

Fig. 1. Anaerobic microbial food chain where cellulose is converted to methane (Wolfe, 1971 [474])

organic polymers are hydrolyzed by extracellular enzymes to sugars which are in turn fermented to a variety of fatty acids, alcohols, carbon dioxide or hydrogen. Acetate, methanol, formate or carbon dioxide and hydrogen are the preferred substrates for methanogenic bacteria. Traces of hydrogen gas are usually found along with carbon dioxide in methane fermentations. In anaerobic waste treatment facilities the reported composition of gas evolved range from 65 to 90% (v/v) methane, 5–35% (v/v) carbon dioxide, 0–10% v/v) hydrogen and small amounts of nitrogen [151]. Similar results were obtained with manure [83] and in rumen gas production where hydrogen gas usually constitutes only 0.05% (v/v) of the total rumen gas production [169] or 3×10^{-4} atm. [170]. Hydrogen is synthesized first and is then replaced by methane in these environments [170, 408].

Studies have shown that pure cultures of methanogenic bacteria use carbon dioxide as an electron acceptor and molecular hydrogen as an electron donor according to Eq. 1 [51]:

$$4\,H_2 + CO_2 \rightarrow CH_4 + 2\,H_2O. \tag{1}$$

Certain other carbohydrates or intermediates are first converted to CO_2 which is reduced to methane if hydrogen gas is present [495]. Evidence to support hydrogen utilization in methane synthesis is supported by three lines of evidence. First, Stephenson and Stickland (1933a) [393], Hungate (1967) [170], and Bryant *et al.* (1968) [51] demonstrated an absolute dependence of CH_4 formation upon the presence of H_2 by the methanogenic organisms. Second, studies with specific inhibitors of methanogenesis [476, 426, 87, 408] in anaerobic digestion demonstrated expected hydrogen gas accumulation in every instance. However, accumulation of hydrogen gas was never as great as expected. Third, the addition of glucose [486, 408] leads to high rates of hydrogen gas and carbon dioxide production within twenty-four hours. The disappearance of hydrogen was followed by the concomitant appearance of methane [408]. The rates of production of methane and the uptake of H_2 from the gas phase were each proportional to the concentration of H_2 in the gas phase [87].

The use of hydrogen as an electron donor is not confined to methanogenic bacteria. Molecular hydrogen can be used as the initial electron donor in energy yielding reactions by species of *Hydrogenomonas* [294]; by *Micrococcus denitrificans* [221] with nitrate as the terminal electron acceptor; by species of *Desulfovibrio* [323] with sulfate as terminal electron acceptor; by *Clostridium aceticum* [463, 281a] which reduces CO_2 with H_2 and metabolically synthesizes acetic acid; by *Azotobacter* [298]; by the purple-sulfur bacteria such as *Rhodospirillum rubrum* [287]; by the green-sulfur bacteria [308] and by algae [199]. A review of the energy-coupling mechanisms associated with the oxidation of H_2 by chemoautrophic bacteria was completed by Peck, 1968 [305]. For a review on energy conservation in chemotrophic anaerobic bacteria, see Thauer *et al.* (1977) [425].

Hydrogen gas production appears to be a major product in the catabolism of carbohydrates and other organic compounds by a wide variety of microorganisms (see reviews by Zobell (1947) [497]; Gest (1954) [123]; Greg and Gest (1965) [133]; Mortenson and Chen (1974) [264], and Zajic and Brosseau (1976) [489]). The biochemical reactions are far more numerous than originally thought. The widespread occurrence of hydrogen gas producing microorganisms of widely different taxonomic and physiological types is summarized in Table 2. This table should not be interpreted as supporting a theme that a significant research effort for the production of hydrogen gas has been made. Almost all past studies are incidental or secondary to some other basic studies such as investigation in the rumen [182] or in the anaerobic digestion of sewage sludge [408].

Table 2. Bacteria reported to evolve molecular Hydrogen

Part 1.	The phototrophic bacteria	
	Family I.	*Rhodospirillaceae*
	Genus I.	*Rhodospirillum rubrum* 119, 120, 123, 286, 287, 125, 39, 308, 261, 28, 216, 354, 364, 433, 357, 358, 359, 359a
	Genus II.	*Rhodopseudomonas palustris* 219
		Rhodopseudomonas capsulata 219, 218, 447, 218, 157, 258
		Rhodopseudomonas gelatinosa 219, 218
		Rhodopseudomonas acidophila 219, 218
	Genus III.	*Rhodomicrobium vannielii* 160, 219, 218
	Family II.	*Chromatiaceae*
	Genus I.	*Chromatium* 8, 28, 216
		Chromatium vinosum 219, 218
		Chromatium minutissimum 219, 218, 216
	Genus V.	*Thiocapsa roseopersicina* 218
	Family III.	*Chlorobiaceae*
	Genus I.	*Chlorobium* 308, 219, 218
	Genus III.	*Chloropseudomonas* 28, 216, 217, 218, 219
	Genus IV.	*Pelodictyon* 308, 218, 219
Part 7.	Gram-negative aerobic rods and cocci	
	Family I.	*Pseudomonadaceae*
	Genus I.	*Pseudomonas* 34
	Family II.	*Azotobacteraceae*
	Genus I.	*Azotobacter (Azomonas) agile* 172
		Azotobacter chroococcum 136
		Azotobacter vinelandii 228, 56
	Family III.	*Rhizobiaceae*
	Genus I.	*Rhizobium leguminosarum* 162, 29, 97, 98, 211
Part 8.	Gram negative facultatively anaerobic rods	
	Family I.	*Enterobacteriaceae*
	Genus I.	*Escherichia coli* 143, 144, 189, 428, 137, 471, 397, 204, 390, 392, 488, 414, 413, 239, 284, 376, 436, 187, 437, 450, 241, 32, 403, 123, 124, 19, 299, 300, 94, 154, 292, 133, 117, 78, 79, 206, 468, 337, 9, 445, 207, 264, 351, 101, 82, 72, 491
	Genus III.	*Citrobacter intermedius* 491, 492, 493
	Genus IV.	*Salmonella enteritidis* 189

Table 2 (continued)

Genus VII. *Enterobacter (Aerobacter) aerogenes* 144, 145, 204, 392, 284, 376, 403, 237
Enterobacter (Aerobacter) sp. 340
(Aerobacter) cloacae 204, 284, 376
(Aerobacter) indologenes 43, 450, 122, 123
Genus IX. *Serratia sp.* 278
Serratia marcescens 306
Serratia plymuthica 55
Genus X. *Proteus vulgaris* 295, 412, 376, 161
Proteus mirabilis 376, 92

Family II. *Vibrionaceae*
Genus II. *Aeromonas sp.* 403, 133

Part 9. Gram negative anaerobic rods

Family I. *Bacteroidaceae*
Genus I. *Bacteroides clostridiiformis* 182
Genus II. *Fusobacterium necrophorum* 182
Genus –. *Desulfovibrio sp.* 484, 53, 425
Desulfovibrio desulfuricans 256, 464, 3
Desulfovibrio vulgaris 353
Genus –. *Butyrivibrio fibrisolvens* 182
Genus –. *Selenomonas ruminantium* 353

Part 11. Gram negative anaerobic cocci
Family I. *Veillonellaceae*
Genus I. *Veillonella alcalescens (gazogenes)* 119, 120, 177, 178, 121, 344, 345, 346, 95
Genus III. *Megasphaera elsdenii (Peptostreptococcus* 441, 11, 12, 182

Part 13. Methane producing bacteria

Family I. *Methanobacteriaceae*
Genus I. *Methanobacterium omelianskii* 19, 179
Methanobacterium omelianskii-S 50, 333, 334
Methanobacterium ruminantium 431, 432

Part 14. Gram positive cocci

Family I. *Micrococcaceae*
Genus I. *Micrococcus aerogenes* 458, 86
Micrococcus lactilyticus (Veillonella alcalescens) 109, 457, 177, 178, 459, 298, 440

Family III. *Peptococcaceae*
Genus I. *Peptococcus aerogenes* 109, 458
Peptococcus glycinophilus 70, 203
Peptococcus prevotii 109
Genus III. *Ruminococcus albus* 167, 168, 173
Ruminococcus flavefaciens 182
Genus IV. *Sarcina maxima* 372, 69, 92
Sarcina ventriculi 204, 372, 69

Table 2 (continued)

Part 15.	Endospore forming rods and cocci	
	Family I.	*Bacillaceae*
	Genus I.	*Bacillus macerans (acetoethylicus)* 386, 94
		Bacillus cloacae (Enterobacter cloacae) 295
		Bacillus macerans 493, 141, 94, 133
		Bacillus polymyxa 326, 386, 159, 131, 94, 403, 365
	Genus III.	*Clostridium acetobutylicum* 180, 47, 94, 281
		Clostridium botulinum 428, 5, 75
		Clostridium butylicum 288, 212, 70, 224, 214, 465, 14, 258, 123, 300
		Clostridium butyricum 399, 375, 204, 193, 47, 123, 94, 2
		Clostridium cellobioparum 72
		Clostridium cellulosolvens 85
		Clostridium dissolvens 201, 5
		Clostridium fossicularum 282, 283, 451, 5
		Clostridium hydrogenicus 451
		Clostridium kluyveri 18, 36, 37, 378, 379, 380, 307, 202, 10, 113, 331, 416, 417, 183, 356, 421, 478, 185, 25, 478, 425
		Clostridium oedematis-maligni 428
		Clostridium pasteurianum 258, 367, 368, 259, 440, 442, 59, 56, 57, 146, 148, 208, 194, 273, 274, 470, 186, 264, 60, 71
		Clostridium sporogenes 150, 63, 163
		Clostridium tetani 471, 76
		Clostridium tetanomorphum 15, 16, 19, 480, 430
		Clostridium thermocellum 450a
		Clostridium thermosaccharolyticum 370, 2
		Clostridium welchii 471, 296
		Clostridium werni 451
Part 16.	Gram positive asporogenous rod-shaped bacteria	
	Family I.	*Lactobacillaceae*
	Genus I.	*Lactobacillus delbrueckii* 239
Part 17.	Actinomycetes and related organisms	
	Family I.	*Propionibacteriaceae*
	Genus II.	*Eubacterium limosum (Butyribacterium rettgeri)* 311, 2, 182,

1. Biological Oxidations

Bacterial growth is concerned with transforming chemical and physical energy into biological energy. This transformation is normally associated with ATP as biological energy which is of use to cellular organisms. Biological oxidation serves to provide useful energy (ATP) for cellular processes and to transform nutrients into cellular constituents. The electrons removed from the substrate can flow through an organized arrangement of electron carriers from that of lowest to that of the highest potential (Fig. 2). The terminal electron acceptor in aerobic organisms is oxygen and the free energy made available in the course of this electron flow can result in newly synthesized ATP. Organic

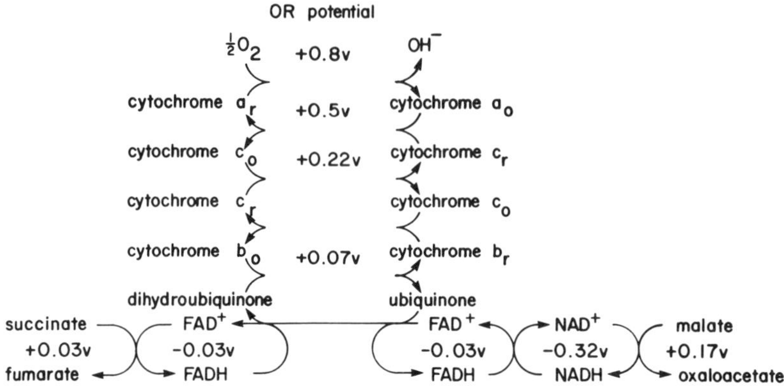

Fig. 2. Apparent organization and electron flow in the mitrochondrial electron transport chains. Approximate values of E_0' are shown (White *et al.*, 1973). r = reduced form, o = oxidized form, FAD = flavin adenine-dinucleotide, NAD = nicotinamide adenine-dinucleotide, FADH = FADH + H^+, NADH = NADH + H^+

or reduced inorganic compounds can serve as electron donors with oxidized inorganic or organic compounds serving as the ultimate electron acceptor [91, 140, 425]. Microorganisms that perform aerobic respiratory metabolism may also be able to grow under anaerobic conditions using metabolic pathways involving anaerobic respiration or fermentation. Some facultative anaerobes can use nitrate as terminal electron acceptor under anaerobic conditions. Anaerobic respiration with sulfate or carbonate as the terminal electron acceptor is normally associated with strictly anaerobic processes [388]. Oxidative, photo or substrate level phosphorylations are methods of useful energy (ATP) formation in bacteria (see review by Haddock and Jones (1977) [140]; Thauer *et al.* (1977) [425]). ATP synthesis is coupled to the oxidation of compounds with transfer of electrons by light-induced or chemotrophic redox processes linked to the reduction of electron acceptors [91, 425]. The functioning of these redox processes are aided by redox components and redox carriers such as cytochromes, quinones, flavodoxins and a various iron-sulfur proteins such as ferredoxin (for reviews on ferredoxin and flavodoxin, see Benemann and Valentine (1971) [23]; Lovenberg (1974) [245]).

Redox carriers are apparently arranged in oxidoreduction loops or membrane segments containing an alternate sequence of hydrogen and electron carriers according to one of the three proposals for the functional organization of ATP production [425, 140]. *E. coli,* for example, can synthesize a variety of redox carriers, depending upon many factors including the growth phase, the terminal electron acceptor, the carbon source for growth, and the strain [140].

A proposed functional organization of the redox carriers responsible for anaerobic electron transport with formate or NADH as electron donor fumarate as terminal electron acceptor in *E. coli* is put forth in Haddock and Jones (1977 [140]. The general features, thermodynamic efficiencies of energy transformation and conservation in chemotrophic anaerobes are treated in Thauer *et al.* (1977) [425]. Fermentation as it is used in this review is looked upon as an anaerobic energy yielding metabolic process

where different organic metabolites usually derived from a fermentable substrate serve as an electron acceptor [388, 425]. Carbohydrates are the principal substrates. In a classical fermentation process, substrate level phosphorylation is regarded as the only mode of ATP synthesis. Exceptions have been reported [425].
The energy obtained by cells (chemotrophic or lithotrophic) is used to drive an endergonic synthesis of ATP from adenosine 5-diphosphate (ADP) and inorganic orthophosphate (P_i). The ATP formed contains an "energy rich" pyrophosphate bond which can be used to perform work, or to drive the production of "energy rich" electrons [23, 285] transported by "energy rich" (low redox) electron carriers [23]. Thauer *et al.* (1977) [425] pointed out that the concept of ATP formation from ADP and P_i and its hydrolytic cleavage to ADP and P_i or AMP and PP_i (pyrophosphate) is a useful formalism to convert complex processes into simple hydrolysis and condensation reactions. Simple hydrolysis in cells would result in the energy being lost as heat (see also Banks and Vernon (1970) [13]). The simple hydrolysis and condensation of ATP is therefore utilized to provide a measure for a specific kind of chemical potential. This chemical potential reflects the coupling between catabolism (bio-energy yielding reactions) and anabolism (bio-energy consuming reactions). In general, approximately -10 to -12 kcal are required for the synthesis of 1 mol of ATP from ADP und P_i in anaerobic bacteria, and approximately 0.1 mol of ATP is required for the synthesis of 1 g cells [21].

$$ADP + P_i \rightarrow ATP + H_2O.$$

Many enzymes involved in ATP synthesis by substrate level phosphorylation (SLP) in anaerobic bacteria catalyze completely reversible processes. These ATP producing systems if and when at equilibrium cannot perform work on their surroundings. For work to be performed one or more reactions of the system must be irreversible [425], such as the ATP-consuming reactions associated with anabolism. Similarly, the anaerobic dehydrogenation of glucose with protons as the elctron acceptor (i.e. H_2 formation) can proceed only to acetate plus CO_2 yielding 49.3 kcal/mol of glucose. Acetate is not normally metabolized under anaerobic conditions by *E. coli* for example, since the dehydrogenation of acetate to CO_2 by the TCA cycle is endergonic requiring 25 kcal/mol The reactions known to be coupled with SLP are summarized in Thauer *et al.* (1977) [425]. In general with one mol of ATP synthesized one mol of protons is formed in addition to a mol of H_2 evolved as a result of substrate dehydrogenation.

2. Substrate and Nutrient Requirements

a) Pyruvate

Pyruvate is a major intermediate in the microbial breakdown of carbohydrates. The clostridia and coli-aerogenes bacteria under anaerobic conditions convert pyruvate to acetic acid and carbon dioxide with the formation of hydrogen gas. This conversion of pyruvic acid occurs only in the presence of a hydrogen acceptor and not under aerobic conditions [239]. The *E. coli* reaction, however, occurs in two distinct stages [436]. In the first stage pyruvate is converted to acetyl phosphate and formic acid.

$$CH_3COCOOH + HPO_4^{2-} \rightarrow CH_3COPO_4^{2-} + HCOOH. \quad (3)$$

The production of acetyl-phosphate and formic acid as a result of the anaerobic metabolism of pyruvate by *Escherichia coli* was termed the "phosphoroclastic" reaction [187, 437, 241]. Subsequent studies described a "thioclastic" rather than a phosphoroclastic reaction catalyzed by pyruvate: formate lyase [403, 292, 207],

$$CH_3COCOOH + HSCoA \rightarrow CH_3COSCoA + HCOOH. \quad (4)$$

The acetyl moiety of pyruvate in the coli-types is transferred to coenzyme A, yielding acetyl-CoA.

A pyruvate dehydrogenase complex and a pyruvate: formate lyase complex are two alternative enzyme systems of *E. coli* that accomplish the transformation of pyruvate into acetyl-CoA [337] and formic acid [206]. The pyruvate-dehydrogenase complex is inoperative under anaerobiosis [142], whereas the pyruvate : formate lyase reaction does not proceed in aerobiosis [154, 206, 207]. Pyruvate : formate lyase appears in aerobically grown cells within a few minutes upon removal of oxygen [154], even if protein synthesis is blocked by chloramphenicol [207].

The gaseous end-products of glucose degradation (or pyruvate dissimilation) by various H_2-producing microorganisms is shown in Table 3. Wide diversity in H_2-producing capability can be seen among these microorganisms. The maximum quantity of hydrogen

Table 3. The gaseous end-products of glucose degradation by some strains of various H_2-producing microorganisms

Organism	Author	Gaseous end products Mol/mol Substrate		Ratio
		H_2	CO_2	
E. coli	144	1.0	1–0.72	1.0
	398	–	–	–
	33	0.003	0.02	0.15
	491	0.15	0.05	3.0
C. intermedius	492	0.85	0.37	2.3
Aerobacter	340	0.36	1.72	0.2
Serratia	278	0.60	1.50	0.4
C. butyricum	94	2.33	1.96	1.18
C. acetobutylicum	94	1.4	2.2	0.64
Lactic acid Bacteria	94	0.74	0.5	1.48
S. maxima	94	2.33	1.49	1.56
B. macerans	94	1.35	2.15	0.63
B. polymyxa	94	0.75	2.03	0.37
R. albus	173	2.6	2.0	1.3
R. albus–V. succinogenes	173	4.0	2.0	2.0

gas reported to evolve per molecule of glucose among the *E. coli* or related microorganisms is one. Greater quantities have been reported to be evolved among the clostridia, bacilli and *Sarcina*.

Approximately 2.6 mols H_2 have been reported formed by *Ruminococcus albus* per mol of hexose [173]. This value (2.6 mol H_2 at 1 atm) may be low due to the inhibitory effects of relatively high hydrogen gas partial pressure [173, 425]. Four mols of H_2 is the highest amount ever reported to have been obtained from hexoses in anaerobiosis [171, 425]. This amount (4 mols H_2) is theoretically equivalent to 33% of the combustible energy of organic compounds [424]. Approximately 2.0 mols CO_2 was evolved per mol hexose in addition to the H_2 formed by *R. albus* [173]. As a result, the H_2/CO_2 ratio (Table 3) was found to be 1.3. Greater ratio values (2.0) were obtained when *R. albus* was grown in coculture with *Vibrio succinogens,* due to increased H_2 evolved. An *E. coli* citrate-utilizing strain isolated from sewage sludge has been observed to evolve H_2 and CO_2 in Durham tubes from glucose [491]. The H_2/CO_2 values were higher than any of those reported in the literature [490, 491], dissolved CO_2 and bicarbonate, however, were not included. Further studies with four other *E. coli* American type cultures gave similar ratios of H_2/CO_2 obtained by the sewage sludge isolate indicating that this capability is quite common to *E. coli* (Brosseau and Zajic, unpublished results). Studies performed with *Citrobacter intermedius* growing on glucose medium under anaerobic conditions with a stirred tank 14-liter fermenter and an atmosphere that was continuously removed and collected resulted in a final gas phase H_2/CO_2 gas ratio of 2.3 (Brosseau and Zajic, unpublished results).

b) Fumarate

In aerobically growing organisms, acetyl-CoA is oxidized to CO_2 via the citric acid cycle [456]. As constituents of this aerobic citric acid cycle succinyl-CoA is formed (α-ketoglutarate dehydrogenase) from α-ketoglutarate and succinate is formed (succinyl-CoA synthetase) from succinyl-CoA along with the production of ATP or GTP and CoA [44]. The formation of succinate from α-ketoglutarate is not believed to occur in anaerobically growing organisms except in *Proteus rettgeri* [223]. Succinate formation can be coupled with acetate and succinyl-CoA to form acetyl-CoA in *P. Rettgeri* [425]. In many anaerobic bacteria fumarate acts as an electron acceptor and is reduced to succinate (see review by Thauer *et al.* (1977) [425]). This is also the case for *E. coli* [155, 247, 140]. Fumarate can be formed from malate or pyruvate plus carbonate and can oxidize various hydrogen donors such as NADH or formate with the formation of succinate and resulting phosphorylation. Fumarate is important because many anaerobic bacteria can carry out fermentations with H_2 as electron donor and either fumarate, sulfate, nitrate or CO_2 as electron acceptor with the concomitant production of ATP [425]. In summary, anaerobically growing microorganisms cannot oxidize acetyl-CoA to CO_2 because the citric acid cycle can function only if succinate can be oxidized to fumarate. This is possible only with electron acceptors with a redox potential more positive than +33 mV (succinate/fumarate; $E_0' = +33$ mV) such as O_2, NO_3^-, NO_2^-, Fe^3 [425]. This makes acetyl-CoA the most frequently used source of high energy in anaerobic microorganisms.

c) Acetyl-CoA

Acetyl-CoA cannot be oxidized to CO_2 via the citric acid cycle in anaerobically growing organisms. The thioester energy bonds of acetyl-CoA formed under anaerobic conditions is conserved as ATP through the intermediate acetyl phosphate [404] by the activity of the enzyme phosphotransacetylase [366];

$$CH_3COSCoA + HPO_4^{2-} + HPO_4^{2-} \rightarrow CH_3COPO_4 + HSCoA \quad (5)$$

and the enzyme acetate kinase [239, 348, 427];

$$CH_3COPO_4 + ADP \rightarrow CH_3COOH + ATP. \quad (6)$$

Phosphotransacetylase and acetate kinase are found in all anaerobic bacteria that form acetyl-CoA to synthesize ATP as well as in a few aerobic bacteria (see review by Thauer *et al.* (1977) [425]).
A comparative scheme showing production of hydrogen gas from the anaerobic breakdown of pyruvate by clostridia and *Escherichia coli* is shown in Fig. 3. Electron carriers of low redox potential and close relationship with pyruvate, an anaerobic glycolytic by-product, function in the formic hydrogenlyase systems [133, 9, 264] of the *coli-aerogenes* group, whereas ferredoxin is an important electron carrier found in both clostridial and *coli*-hydrogen gas producing systems. Pyruvate: ferredoxin oxidoreduc-

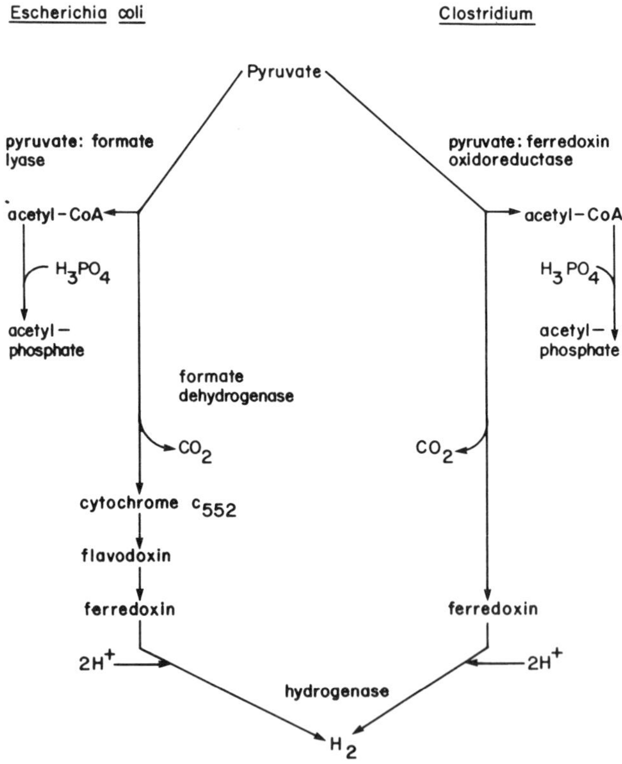

Fig. 3. Comparative scheme showing production of hydrogen gas from the anaerobic breakdown of pyruvate by saccharolytic clostridia and *Escherichia coli* (after Mortenson and Chen, 1974 [264])

tase in the clostridial reaction is the enzyme that catalyzes pyruvate oxidation yielding acetyl-CoA and CO_2 in the presence of the electron acceptor ferredoxin [438, 439]. In the absence of ferredoxin or some other electron acceptor, the enzyme is reduced by pyruvate only in the presence of coenzyme A (see Fig. 4). Stoichiometric amounts of $^{14}CO_2$ are released from (1 – ^{14}C) pyruvate even in the absence of coenzyme A.

Clostridial system
(1) pyruvate + TPP-E_0 = hydroxyethyl-TPP-E_0 + CO_2
(2) hydroyethyl-TPP-E_0 + CoASH = acetyl-SCoA + TPP-E_r
(3) TPP-E_r + Fd_0 = TPP-E_0 + Fd_r
(4) acetyl-CoA + P_i = acetyl-phosphate + CoASH
(5) Fd_r + 2 H^+ = Fd_0 + H_2

Escherichia coli system
(1) pyruvate + E = acetyl-E + formate
(2) acetyl-E + CoA = E + acetyl-CoA
(3) acetyl-CoA + phosphate = acetyl-phosphate + CoA
(4) HCOOH + Fd_0 = CO_2 + Fd_r
(5) Fd_r + 2 H^+ = Fd_0 + H_2

Fig. 4. Biochemistry of the production of hydrogen gas in the clostridial and *Escherichia coli* systems. The clostridial system is catalyzed by pyruvate: ferredoxin oxidoreductase [E] (1–3), phosphotransacetylase (4), and hydrogenase (5). The *coli*-system is catalyzed by pyruvate:formate lyase [E] (1–2), phosphotransacetylase (3) and formic hydrogenlyase (4–5). Fd = ferredoxin, E = enzyme, o = oxidized form, r = reduced form, TPP = thiamine pyrophosphate

Pyruvate first combines with the coenzyme thiamine pyrophosphate (TPP) and is decarboxylated. This yields an intermediate, hydroxyethyl-TPP and carbon dioxide. The thiamine pyrophosphate is regenerated for reuse by the transfer of the hydroxyethyl group to coenzyme A (CoASH) and the reduction of oxidized ferredoxin. Those organisms that also contain phosphotransacetylase and inorganic phosphate can generate acetyl-phosphate and CoASH. Reduced ferredoxin is oxidized to oxidized ferredoxin and hydrogen gas in the presence of hydrogenase. The high energy acetyl-phosphate bond can be transferred to ADP with ATP formation.

In *Clostridium pasteurianum* NADH is primarily formed by an NAD-glyceraldehyde phosphate dehydrogenase and reduced ferredoxin is obtained from pyruvate dehydrogenation to acetyl-CoA and CO_2 [186]. The primary source of reducing equivalents for H_2 formation is obtained from pyruvate dehydrogenation, whereas much of the NADH produced is needed for butyrate production. Butyrate is produced as a result of the condensation of two mols of acetyl-CoA forming acetoacetyl-CoA which is reduced to butyryl-CoA, and converted to butyrate via phosphotransbutyrylase and butyrate kinase.

Some of the NADH is made available for H_2 formation by way of ferredoxin because more H_2 is produced than pyruvate is oxidized [417, 186, 425]. Hydrogen gas formation is dependent upon the presence and activity of NADH : ferredoxin oxidoreductase which requires acetyl-CoA as an allosteric activator although CoA is completely antagonistic. The acetyl-CoA/CoA ratio therefore regulates the quantity of H_2 evolved as well as the ATP generated in the acetate kinase reaction. This acetyl-CoA/CoA ratio must also control that quantity of acetyl-CoA converted into butyric acid. Hydrogen gas production among strict anaerobic bacteria, however, appears to be associated with butyric acid production. Pure cultures of strict anaerobic bacteria isolated from primary sewage sludge that produced hydrogen also produced butyric acid, those that did not produce hydrogen did not produce butyric acid [409].

The "*coli*-type" system associated with facultative anaerobic microorganisms is similar to the clostridial system except that pyruvate dissimilation does not yield free-CO_2. The electrons removed by the reaction are transferred not to protons but to the CO_2 produced in the decarboxylation of pyruvate (Fig. 4). The ability to dissimilate pyruvate into acetyl-CoA and formic acid was found to arise only from interaction between pyruvate : formate lyase, a ferrous ion-activated enzyme, S-adenosylmethionine and an appropriate redox potential where thiamine diphosphate may play a role [206]. Thiamine diphosphate however is not a constituent of pyruvate : formate lyase [206, 207]. Although cytochrome C_{552} has been considered to be that soluble factor of low redox potential participating in the hydrogen gas evolution reaction [351, 78, 124, 299, 300, 9, 132], results of recent experiments [101] indicate that cytochrome C_{552} is not required for formate hydrogen lyase activity. It is instead considered to be a component of a NADH:nitrite oxidoreductase catalyzing the reduction of nitrite [82, 79].

d) Formate

According to a review article by Zobell (1947) [497], Popoff (1875) was the first to observe evolution of molecular hydrogen by bacterial action. It was noticed that a mixed culture fermentation of calcium formate from pond mud was capable of producing hydrogen gas but not methane. Hoppe-Seyler (1876) continued the work and established the relationship of hydrogen gas and carbon dioxide evolved from calcium formate as:

$$Ca(OOCH)_2 + H_2O \leftrightharpoons CaCO_3 + CO_2 + 2\,H_2. \tag{7}$$

Van Tieghem (1877) and Prazmowski (1880) described the fermentation of cellulose by ***Bacillus amylobacter, Bacillus polymyxa*** and ***Vibrio rugola*** with the formation of hydrogen gas, carbon dioxide and organic acids [497]. Pure cultures of various bacterial species were used in the production of gas from formate, glucose and other carbohydrates [110–112, 295] The gas produced was a mixture consisting of hydrogen and carbon dioxide in a ratio of 1:1. Formic acid was also accumulated in large quantities when the evolution of gas was stopped by allowing the gas pressure to go too high. Among the gas producing microorganisms were: *B. lactis aerogenes, B. cloacae, B. coli communis, B. ethaceticus* and *Proteus vulgaris.* In the same year Harden (1901) [143] confirmed that an increase of H_2 partial pressure increased the yield of formic acid from glucose and inhibited further decomposition of this acid.

The mechanism of hydrogen gas production involving formic acid was elucidated by Stephenson and Stickland (1932) [392]. The formic acid formed as a result of the pyruvate : formate lyase in *coli*-type microorganisms is oxidized to CO_2 with the formation of hydrogen gas [392];

$$HCOOH \leftrightharpoons CO_2 + H_2. \tag{8}$$

Similar reactions that convert pyruvic acid into acetic acid, CO_2 and H_2 in clostridial species [480, 212–214, 241] fail to act on formic acid.
In *E. coli* formic acid synthesis is independent of formate dehydrogenase and is formed from pyruvate by pyruvate formate lyase. Formic acid produced is oxidized by the formic hydrogenlyase enzyme complex to CO_2 with the formation of H_2 (Fig. 4). The reactions catalyzed by the hydrogenase and formic hydrogenlyase are readily reversible and can be inhibited by the presence of certain quantities of hydrogen gas [434]. Hydrogen gas also inhibits the growth of the hydrogen gas producing *Clostridium cellobioparum* but not *E. coli* [72]. A reversible thioclastic [10, 331] or pyruvate : formate-lyase reaction [421, 478] was observed in *Clostridium kluyveri* and *C. butyricum.* The net synthesis of pyruvate occurs from CO_2, acetyl-CoA and reduced ferredoxin, S-adenosylmethionine and a formic acid synthesizing enzyme [184]. The exact mechanism of this reversed pyruvate formate-lyase reaction in *E. coli* serves mainly to mediate acetyl-CoA synthesis for ATP generation in catabolism. The clostridial lyase functions mainly to furnish formate for anabolism. Both require S-adenosylmethionine and a reducing system with flavodoxin or ferredoxin as the physiological reactants.
It was suggested that the reduction of CO_2 to formic acid that occurs in *Clostridium aceticum* is catalyzed by a NADPH-dependent formic dehydrogenase [233, 420]. Ferredoxin also mediates the transfer of electrons to pyridine nucleotides [261]. The presence of a pool of formic acid lowers the incorporation of CO_2 into the methyl group of acetate in *C. aceticum* [243]. This suggests that formate dehydrogenase may be active in clostridia. The historical inability of clostridia to produce hydrogen gas from formic acid in contrast to *E. coli* indicates that formate dehydrogenase is not involved with H_2 production [224, 213]. Recent evidence correlates the activity of reduced ferredoxin: CO_2 oxidoreductase with the accumulation of formic acid in *Clostridium pasteurianum* [422, 423]. The function of the enzyme is to mediate formic acid synthesis rather than formic acid oxidation.
Further studies indicate a formic hydrogenlyase active in the absence of any added iron or sulfur to the growth medium (see also section on sulfur). Although the *E. coli* isolated cannot grow on formate alone, there are other microorganisms which can (see reviews by Pine (1971) [312]; Thauer *et al.* (1977) [425]). These are methanogenic bacteria which presumably contain a formic hydrogen-lyase which catalyzes reaction (7) and a methane fermentation of H_2.

e) Inorganic Nitrogen

The effects of nitrate on hydrogenase and formic hydrogenlyase activities are varied. Hydrogenase activity with oxygen or nitrate as terminal electron acceptors was considered by Gest (1954) [123] to constitute a major energy yielding reaction with respect to growth in autotrophic pseudomonads. This view was not held for hydrogenase containing heterotrophic microorganisms. Two main types of nitrate reduction occur in microorganisms, assimilatory and dissmilatory nitrate reduction [275]. Microorganisms that use nitrate as a source of cell nitrogen do so by assimilatory nitrate reduction. This means that nitrate must be reduced to ammonia. Respiratory nitrate reductases are not inhibited by ammonium salts [467]. The first step in assimilatory nitrate reduction is mediated by nitrate reductase, a flavoprotein containing molybdenum [192]. The enzymes of nitrate assimilation are distinct from cytochrome enzymes involved in dissimilatory (respiratory) nitrate reduction [210]. The respiratory enzyme also called nitrate reductase is a complex consisting of formic dehydrogenase, cytochrome b_1, and nitrate reductase [400, 92, 246].

The reduction of O_2 to H_2O, CO_2 to CH_4, SO_4^{2-} to H_2S, NO_3^- to NO_2^- or N_2 and fumarate to succinate is apparently associated with ATP formation, whereas the reduction of H^+ to H_2 is not [425]. Mechanisms of electron transport and substrate level phosphorylation are reviewed in Thauer *et al.* (1977) [425]. For example, the chemiosmotic hypothesis assumes that a proton-motive force consisting of a pH gradient (ΔpH) and an electrical potential difference ($\Delta\psi$) is generated by redox reactions of electron transport. It is this proton-motive force (translocation) that drives the synthesis of ATP in conjunction with functioning membranes. Facultative anaerobes such as *E. coli* that use oxygen and nitrate as terminal electron acceptors under aerobic and anaerobic conditions respective, do so as a mechanism for handling the removal of excess electrons. These excess electrons have been shown to be associated with proton translocation and the reduction of nitrate to nitrite in *E. coli* [139]. The proton-motive force can be composed solely of a membrane potential or solely of a pH gradient [429].

Under anaerobic conditions nitrates are reduced to nitrite by nitrate reductase [210],

$$NO_3^- + 2\,e^- + 2\,H^+ \rightarrow NO_2^- + H_2O.$$

Recent reviews on nitrate respiration are published by Stouthamer (1976) [401]; Haddock and Jones (1977) [140]; Thauer *et al.* (1977) [425]. One electron donor for the nitrate reductase in *E. coli* is formate which is dehydrogenated by formic dehydrogenase. This formate dehydrogenase is a molybdoprotein containing selenium [105, 107, 231, 369]. The formate dehydrogenase associated with the formic hydrogenlyase system is dissimilar to the enzyme associated with nitrate reduction [350, 351, 92, 93]. Electrons are transferred to a flavoprotein then to cytochrome b_{555}, the immediate donor for nitrate reductase [102, 92]. The pathway of nitrate reduction in *E. coli* is [350],

$$HCOOH \rightarrow F_p \rightarrow \text{Cyt } b'_{555} \rightarrow \text{Cyt } b''_{555} \rightarrow NR \rightarrow NO_3^- \qquad (9)$$

where F_p = flavoprotein, cyt b = b-type cytochrome and NR = nitrate reductase. Another electron donor is NADH. The relationship of the NADH-nitrate reductase system to the normal respiratory chain remains to be solved [350]. Ubiquinone is also specifically involved in nitrate reduction (NO_3^-/NO_2^-; $E_0' = +433$ mV) in gram negative bacteria.

No reduction of NAD or NADP occurred as a result of the reduction of nitrate to nitrite due to the formate dehydrogenase reaction in *Mycobacterium phlei* [96].

The nitrate reductase system is not considered an alternative mechanism for oxidation of reduced NAD [455]. Formate-nitrate reductase pathway is distinct from the rest of the electron-transport system in cells in the log phase of growth although NADH-nitrate reductase acitivity found in stationary-phase cells may be part of the complete electron transport system [455].

Various other substances also serve as terminal electron acceptors under anaerobic conditions. The simplest are protons, that are reduced by hydrogenase of the hydrogenlyase complex resulting in the evolution of hydrogen gas [264]. The general implication of the respiratory nitrate reductase pathway is that the hydrogenase activity in *E. coli* is not necessarily apparent when grown in the presence of nitrate [310, 80, 468], hydrogen gas evolution does not occur [295, 123, 392, 488, 491] and the formate-nitrate system should inhibit hydrogenase activity by the preferential removal of excess electrons. The decrease in hydrogen gas evolution was found to be independent of cell growth on glucose but not so on yeast extract plus formic acid [491]. However, a decrease in cell dry weight as a result of growth on glucose was observed in the presence of 0.1% w/v sodium nitrate or greater, perhaps due to the accumulation of toxic amounts of sodium nitrite in the growth medium. Figure 5 shows the effect of varying concentrations of inorganic salts on the yield of H_2 production per unit cell dry weight yield in the presence of glucose or yeast extract plus formic acid. The presence of nitrate has a different effect on certain algal symbionts which show nitrogenase-catalyzed hydrogen gas evolution. Neither ammonium nor nitrate inhibits nitrogenase catalyzed hydrogen gas evolution but ammonium does repress nitrogenase biosynthesis [279]. Figure 5 also shows the effect of varying concentrations of ammonium chloride on the yield of H_2 production by *E. coli* [490, 491]. H_2 evolution increased proportionately with increases in the ammonium present up to 0.1% w/v (0.02 M) ammonium chloride. A sharp decrease was then observed which may be due to the influence of an active hydrogenase that is active in removing hydrogen gas or involved in the reductive amination of organic compounds [261] Concentrations of 0.1 M NH_4Cl inhibited hydrogenlyase activity in formate plus peptone but not in glucose plus amino acid mixtures [226].

Studies have suggested that the presence of adequate levels of casein hydrolysate inhibit the effect of nitrate on the appearance of hydrogenase and hydrogenlyase activities [31]. Zajic and Brosseau (1976) [491] have shown an inhibitory effect of increasing concentrations of $NaNO_3$ on H_2 gas formation with *E. coli* grown either on yeast extract plus formic acid or glucose. Nearly identical inhibitory curves were obtained (Fig. 5). Hydrogen evolution stops when 0.2% w/v $NaNO_3$ or greater is present in the growth medium. There are strains of bacteria which hydrolyze proteins and amino acids, and seem unable to attack carbohydrates. Several are able to grow at the expense of one single amino acid with the formation of a series of end products including hydrogen gas [19, 94]. These bacteria include *Escherichia* [123, 491], *Proteus* [412]; *Clostridium* [16, 17, 481, 76, 448]: *Peptococcus* [70, 109, 203]: *Micrococcus* [457, 458, 440]: and *Rhodospirillum* [119, 120]. All amino acids yielding H_2 and/or CO_2 were deaminated. Fatty acids were also produced. The type of fatty acids produced depended on the physiology of the microorganism and the type of amino acid being metabolized. The predominant organic acids produced included lactic, butyric, propionic, acetic and formic. Some purines were also utilized and resulted in similar organic acid end products. Those microorganisms that utilized cystine also produced hydrogen sulfide as an end product. Amino acids utilized in the production of H_2 by microorganisms reported in the literature include cystine, glutamic acid, aspartic acid, serine, glycine, theronine, histidine, as well as purines: hypoxanthine, xanthine and adenine.

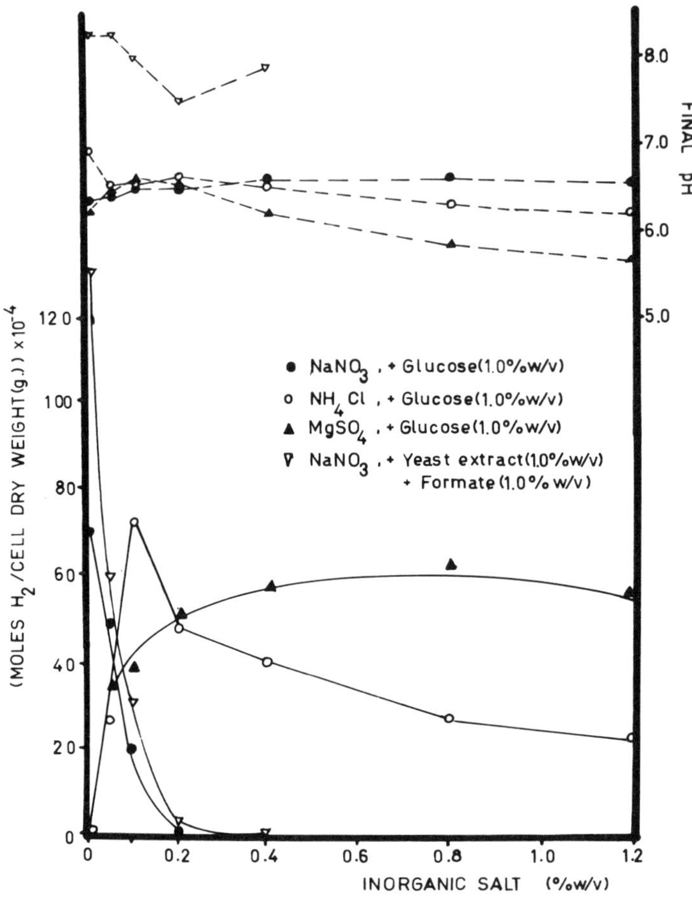

Fig. 5. The effect of varying concentrations of $NaNO_3$, NH_4Cl or $MgSO_4$ as sole sources of nitrate, cellular nitrogen and cellular sulfur on the final yield of H_2 with media containing glucose. Nitrate studies were also performed with media containing yeast extract plus formic acid (Zajic and Brosseau, 1977 [490, 491]). Experiments were performed with Durham tubes. The original pH = 6.8

The *E. coli* isolated by Zajic and Brosseau (1976) [491], produces little or no gas on yeast extract or beef extracts alone. Upon addition of formic acid approximately 3–4 times the quantity of H_2 is evolved, as compared with growth upon equal quantities of glucose. The addition of formic acid to a glucose mineral salts medium enhanced H_2 production with growth. Hydrogen evolution stopped in the presence of concentrations greater than 0.7% w/v sodium formate (Zajic and Brosseau, unpublished results). Vitamin-free casamino acids, nutrient broth and polypeptone were able to replace yeast extract and support H_2 evolution with the addition of formic acid to the growth medium. Casein and peptone were less able to do so. No gas evolved as a result of growth upon acetate, propionate, butyrate, or yeast extract plus acetate.

f) Inorganic Sulfur

Most organisms can utilize sulfate reducing it to sulfide and incorporate the sulfide into cellular organic materials. This type of sulfate reduction has been termed assimilatory sulfate reduction [303, 230]. Assimilatory sulfate reducers can also reduce inorganic sulfur compounds other than sulfate (Fuchs and Bonde (1957) [116]. Sulfate utilized is reduced to sulfide where an organic compound such as serine acts as the sulfide acceptor [230]. The conversion of sulfate to sulfide is an eight electron transfer process:

$$4\,H_2 + SO_4^{2-} \rightarrow S^{2-} + 4\,H_2O. \tag{10}$$

The equivalent of four mols molecular hydrogen are required to reduce one mol of sulfate. The effect that certain organic and inorganic sulfur compounds have on the final yields of H_2 with *C. intermedius* have been studied [492]; Brosseau and Zajic (1977), unpublished results). The addition of $MgSO_4$ to the growth medium of *C. intermedius* as a sole source of sulfur has a stimulatory effect on H_2 evolution [490, 491, 492]. Preliminary results with Durham fermentation tubes indicate that the addition of either of the following compounds: $MgSO_4$; K_2SO_4; $(NH_4)_2SO_4$; Na_2SO_3; $Na_2S_2O_3$; Na_2S to the growth medium of *C. intermedius* as sole sources of sulfur stimulate gas evolution [492]. Indications are that a form of utilizable sulfur more reduced than equivalent amounts of $MgSO_4$ or K_2SO_4 in the growth medium result in greater yields of H_2. Gas composition ranged between 75–78% (v/v) of the gas produced with CO_2 constituting the remainder of the gas. H_2S has not been detected in the growth medium of *C. intermedius;* neither H_2S nor ammonia were detected by gas chromatography among the gases evolved. H_2 accumulation was prevented with media containing inhibitory concentrations of sulfur salt such as sodium thiosulfate, sulfite, or sulfide. The pattern of gas yields obtained indicated that larger relative yields of H_2 resulted with thiosulfate or sulfide (0.21 mols H_2/mol glucose). The lowest optimal yields of H_2 obtained per gram of biomass were achieved with $MgSO_4$. This is thought to reflect greater cellular yields obtained with this compound. It has been determined that approximately 20% of the total hydrogen evolved under batch growth conditions is collected in the Durham tubes (Brosseau and Zajic, 1977, unpublished results). Even so, consistent and significant relative increases in yields of H_2 were detected. Studies are presently underway to evaluate the Durham tube results using a stirred tank 14-liter fermenter with strict anaerobic conditions and *C. intermedius.* Recent results indicate maximal yields of 14.3 mmols $H_2 \cdot (g)\ \text{biomass}^{-1} \cdot \text{hr}^{-1}$ or 314 ml $H_2 \cdot (g)\ \text{biomass}^{-1} \cdot \text{hr}^{-1}$ with growth on glucose (Brosseau and Zajic).

The use of sulfate as a respiratory electron acceptor is termed dissimilatory sulfate reduction:

$$SO_4^{2-} + 8\,e^- + 8\,H^+ \rightarrow S^{2-} + 4\,H_2O. \tag{11}$$

Some genera (*Desulfovibrio* and *Desulfotomacculum*) [69, 325] are capable of reducing sulfate to sulfide at the expense of molecular hydrogen when suitable sources of carbon are available [391, 302, 304, 251, 323]. The typical habitats of the dissimilatory sulfate reducers are anaerobic sediments which contain organic matter and sulfate [303, 230]. Large quantities of H_2S are generated in these environments, part of an anaerobic sulfur cycle. Sulfite, thiosulfate, tetrathionate or elemental sulfur can also replace sulfate as respiratory electron acceptor [317, 30].

Sulfate reducers were suggested as being instrumental in preventing the accumulation of hydrogen gas resulting from the degradation of organic matter [361]. The reduction of sulfate to sulfide with H_2 as the electron donor and its relationship to the corrosion of iron was described in detail by Starkey (1947) [389]. Evidence indicates that compounds such as sulfite, thiosulfate in addition to sulfate are reduced with hydrogen gas to form hydrogen sulfide [302, 304, 251, 323, 230] and that both ferredoxin and flavodoxin can couple reduction of thiosulfate by molecular hydrogen in *Desulfovibrio* [138, 229, 20].

Normally, dissimilatory sulfate reducing bacteria do not utilize carbohydrates and a complete functional tri-carboxylic acid cycle does not exist [129, 232]. The previous autotrophic status of *Desulfovibrio* [64, 65] has been disputed [251, 320, 323]. The conclusion was that CO_2 was assimilated only during heterotrophic growth. As a consequence the principal substrates which serve as both carbon and energy sources are lactate, ethanol, and pyruvate. These and other compounds are oxidized only to the level of acetate [230]. Evidence was presented for oxidative phosphorylation during the reduction of sulfate with hydrogen gas by *D. desulfuricans* and that growth could occur by oxidation of molecular hydrogen or formate [374]. Evidence of phosphorylation was provided by Peck (1966) [304] with cell free preparations of *D. gigas* grown on lactate plus sulfate. Sulfate was reduced to sulfide with H_2 as electron donor and concomitant esterification of phosphate. Previous findings [302] also indicate that dissimilatory sulfate reduction is coupled with phosphorylation (see review by Thauer *et al.* (1977) [425]). Other reports indicate that hydrogen gas was not involved as an energy source with sulfate-reducing bacteria growing on lactate plus sulfate [318, 52, 200]. Growth and hydrogen uptake studies by Khosrovi *et al.* (1971) [200] with a strain of *D. vulgaris* indicated sulfate reduction was in part non-growth associated. H_2 did not serve as an alternative source of energy. *D. vulgaris* and *D. desulfuricans,* however do not utilize or metabolize lactate or ethanol in the absence of sulfate. They will grow on lactate or ethanol alone in the presence of a methane forming bacterium [MOH organism] [52]. It must also be noted that a hydrogenase-free strain of *D. desulfuricans* has been reported to grow normally on lactate or pyruvate in the presence of sulfate [318]. The primary physiological function of hydrogenase may be associated with hydrogen evolution rather than hydrogen utilization proved that the partial pressure of H_2 is kept relatively low. It thus appears that an electron sink in the form of sulfate, protons or the presence of hydrogen gas-utilizing microorgansims may be required to enable the sulfate reducers to maintain a low H_2 partial pressure and prevent the accumulation of growth inhibitory levels of H_2. It is noteworthy that high concentrations of H_2S (1000–2500 μg/l) apparently do not affect hydrogenase activity [254].

Growth of *D. desulfuricans* has been reported to occur on pyruvate in the absence of sulfate [322]. Extracts of *D. desulfuricans* catalyze the decomposition of pyruvate to acetate, CO_2 and H_2 [256]. Similarly H_2 is obtained from formate [464]. Formate is also utilized as an electron donor for sulfate reduction [323].

Cytochrome c_3 (E_0' = –205 mV [321]; –300 mV [485]) is found in this microorganism as a coenzyme of hydrogenase [482] and formate dehydrogenase [483]. This cytochrome was described as fulfilling electron transport functions in *Desulfovibrio* [323, 175, 176, 464, 482]. Yagi (1970) [484] purified hydrogenase and found cytochrome c_3 was functional in the elctron transport involved in H_2 evolution and consumption reactions. A sequence of electron transfer reactions from pyruvate was described by Akagi, 1967 [3] and is shown in Fig. 6. In the oxidation of pyruvic acid by *D. desulfuricans* to ace-

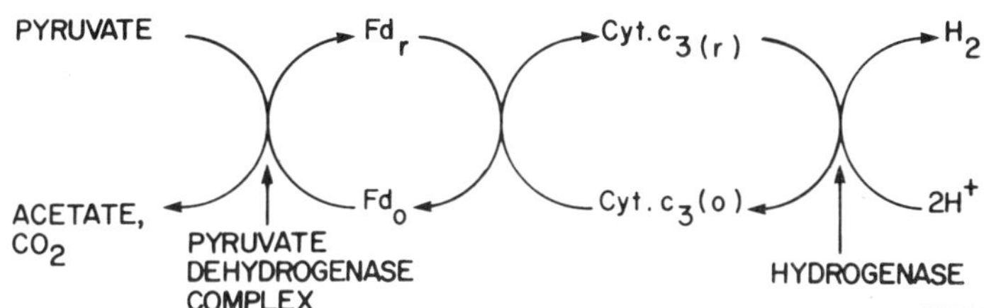

Fig. 6. The sequence of electron transfer from pyruvate as described by Akagi (1967) [3] for *Desulfovibrio*

tate, CO_2 and H_2, ferredoxin accepts the electrons from a pyruvate dehydrogenase (pyruvate : ferredoxin oxidoreductase) complex, whereas cytochrome c_3 does not. However, both ferredoxin and cytochrome c_3 are capable of donating electrons to the hydrogenase of *D. desulfovibrio.* The two carriers in combination result in a maximum production of hydrogen gas and acetyl phosphate by the clastic reaction which takes place. Formate dehydrogenase, hydrogenase as well as ferredoxin, cytochrome c_3 and flavodoxin have been shown to be involved in sulfate reduction to sulfide with H_2 [425]. It is possible that the presence and location of hydrogenase enables anaerobic sulfate respiratory reducers to metabolize substrates that in symbiotic-like growth with other bacteria could not be utilized in pure culture [22].

Ethanol was shown to be catabolized to acetate and H_2 by the nonmethanogenic bacterium *Methanobacterium omelianski* S when cocultured with the methanogenic bacterium *M. omelianski* MOH which produced methane via reduction of CO_2 with H_2 (Bryant *et al.* 1967 [50]). Similar results were obtained by members of the genus *Desulfovibrio* and *M. formicicum* [53]. The conversion of ethanol to acetate and H_2 in pure culture is not thermodynamically favorable since the free-energy change is not negative enough to allow the conversion to proceed unless the partial pressure of H_2 is maintained at a level lower than standard conditions [53, 425]. Growth rates and yields of desulfovibrios on ethanol, lactate or pyruvate were higher with sulfate as electron acceptor than with protons with subsequent use by methanogenic bacteria [53]. Sulfate when in excess inhibits methanogenesis in mixed culture systems [462, 53]. It is suggested that the presence of sulfate eliminates the production of H_2 by desulfovibrios which must also preferentially utilize H_2 produced by other microorganisms for the reduction of sulfate [53]. It is also suggested that acetate would be completely oxidized to CO_2 rather than to methane and CO_2 in the presence of sulfate in anaerobic mixed culture environments [462]. *Desulfotomaculum acetoxidans* which oxidizes acetate to CO_2 and reduces elemental sulfur to sulfide [309] grows with a faster growth rate in medium with acetate-sulfate as the energy source than do methane bacteria using acetate [250, 462, 53]. Conversion of acetate + SO_4 to CO_2 is thermodynamically more favourable than conversion of acetate to CO_2 and CH_4 [425]. Competition for available acetate and H_2 is suggested to be the mechanism by which sulfate inhibits methanogenesis in mixed microbial systems [469].

g) Organic Acids, Fatty Acids, and Lipids

Carbohydrates and the related alcohols and organic acids can all be fermented in such a way as to yield hydrogen gas [477]. The assimilation of acetate and other two carbon compounds as the sole source of carbon has been studied mainly in the enteric bacteria and pseudomonads [220, 373]. Acetic acid formation is accompanied by a depression in H_2 production [253] by Aerobacter, and no gas was evolved during growth of *E. coli* upon acetate, propionate and butyrate media [490, 491, 493]. Lichstein and Boyd, 1951, 1952 [234, 235, 236] implicated oleic acid and other long chain fatty acids in formate breakdown by *coli-aerogenes* bacteria causing a striking and immediate stimulation of hydrogenase acitivity.

Some anaerobic bacteria can oxidize butyric acid and other saturated fatty acids to acetate with protons as the electron acceptor. Thauer *et al.* [425] point out that although such a reaction is considered to be an endergonic reaction under standard conditions, with a hydrogen partial pressure less than 1 atm the reaction could proceed. In addition, this formation of acetate is assumed not to be coupled with phosphorylation. *Clostridium kluyveri* utilized acetate in conjunction with ethanol to produce hydrogen gas among other products. Long chain fatty acids especially of the unsaturated types have been reported to be inhibitory to bacteria [277]. Pure cultures of strict anaerobes that will degrade long chain fatty acids and lipids with hydrogen gas production have not yet been reported.

3. Enzymes

a) Role of Enzymes

Reference to enzymes catalyzing the transfer of hydrogen either to oxygen or to some other acceptor have been reported in various animal or plant tissues and bacteria [497, 123]. Stickland (1929) [397] demonstrated the ability of *Bacillus coli* to anaerobically liberate hydrogen gas from formic acid. This was in contrast to formic dehydrogenase of *Bacterium typhosum* shown by Pakes and Jollyman (1901) [295], Quastel and Whetham (1925) [329], and Stickland (1929) [397] to produce no hydrogen from formate

$$HCOOH + X \leftrightarrows XH_2 + CO_2 \qquad (12)$$

where X represents oxygen or any other acceptor. For example, the enzyme formic dehydrogenase reduces methylene blue in the presence of formate but does not liberate hydrogen gas under aerobic or anaerobic conditions [397]. On the other hand, the enzyme hydrogenase [390, 391] catalyzes the oxidation of molecular hydrogen to yield protons and electrons:

$$H_2 \leftrightarrows 2\,H^+ + 2\,e^-. \tag{13}$$

Stephenson and Stickland (1932) [392] described a third bacterial enzyme called formic hydrogenlyase catalyzing the reaction:

$$HCOOH \leftrightarrows H_2 + CO_2. \tag{14}$$

The hydrogen produced is liberated rather than released in the form of protons and electrons.
The formic hydrogenlyase reaction was investigated [284, 124, 299, 117] where evidence was presented that formate decomposition to hydrogen gas and carbon dioxide is catalyzed by a multi-enzyme system consisting of a formic dehydrogenase, hydrogenase and one or two intermediate factors involved in electron transport. Attempts to isolate and purify components of the complex, especially formic dehydrogenase, have met with only limited success [190]. This also suggests that formic dehydrogenase and hydrogenase are part of a complex, and are possibly combined with an electron carrier system. Electron carriers of low redox potential and close relationship with the anaerobic metabolism of an important intermediate were proposed to function in the formic hydrogenlyase systems. The intermediate involved was pyruvate a product of the anaerobic glycolytic pathway [133, 9, 264].

b) Formic Hydrogenlyase and Formic Dehydrogenase

Induced biosynthesis of formic hydrogenlyase in the *coli-aerogenes* bacteria is dependent upon an energy source, and an external supply of amino acids [32, 226, 314, 435]. In addition to anaerobic conditions, formic acid is required as an inducer [392, 393, 394]. In the absence of added amino acids, induction of formic hydrogenlyase was achieved by growing *Escherichia coli* in a reaction mixture containing glucose, formate and phosphate. Adequate amounts of iron salts had to be present [117]. In the absence of iron, glucose was fermented and organic acids were produced and no formic hydrogenlyase activity was observed, confirming an earlier report [450]. In the absence of iron salts hydrogen gas is not produced and large amounts of formic and lactic acids accumulated. Iron deficient cells were devoid of hydrogenase and formic hydrogenlyase but showed formic dehydrogenase activity when adequate amounts of selenium and moybdenum were present in the growth medium. This confirmed the work of Pinsent (1954) [313]. Formic hydrogenlyase activity was also absent in cultures grown with vigorous aeration [123]. Biosynthesis of the enzyme system is repressed by oxygen [310]. Gray *et al.* (1963) [132] associated this loss of activity to a soluble c-type cytochrome which was formed by anaerobically grown cells of *E. coli* and various *Enterobacteriaceae*
It has also been suggested that oxygen repression in *E. coli* could be via a repressor acting on cytochrome c formation [466].

Formic dehydrogenase activity in *E. coli* under aerobic growth conditions oxidizes formate with oxygen to carbon dioxide and water,

$$HCOOH + 1/2\ O_2 \rightarrow CO_2 + H_2O. \tag{15}$$

Under anaerobic conditions formic dehydrogenase activity is measured by reduction of methylene blue. No gaseous H_2 is observed from formate [434]:

$$HCOOH + Mb \rightarrow CO_2 + Mb \cdot 2\ H. \tag{16}$$

Benzyl viologen ($E_0' = -0.359$ v), a dye accepting single electrons, can also be reduced by formic dehydrogenase under anaerobic conditions.

Formic dehydrogenase has been found in a wide variety of bacterial species. The enzyme is an integral part of the formic hydrogenlyase systems of *E. coli* [299], *Rhodopseudomonas palustrus* [328], and *Aerobacter aerogenes* [237]. Formic dehydrogenase has been isolated [181] from two different Pseudomonads grown on methanol. This enzyme has also been reported in *Methanobacterium omelianskii* [45] and in *Clostridium pasteurianum* [443]. Formate metabolism was shown to be coupled to NAD reduction by way of ferredoxin. The purification of formic dehydrogenase in *Clostridium acidiurici* and other organisms has had only limited success [190]. A soluble cytochrome b_1-linked formic dehydrogenase has been obtained from *E. coli* [350]. The formic dehydrogenase isolated from *Clostridium thermoaceticum* [233] is linked to NADP reduction, whereas enzyme activity of *C. formicoaceticum* [243] and *M. omelianskii* S-organism [474] are linked to NAD reduction via ferredoxin. Tzeng *et al.* (1975) [431, 432] demonstrated an NADP-linked factor F_{420}-dependent formic dehydrogenase activity in *M. ruminantium.*

Studies with *E. coli* by Thauer *et al.* (1974) [422] confirmed earlier reports [105, 369] on the effects of selenium compounds on formic dehydrogenase activity. The formic dehydrogenase activity was increased by the addition of selenium to the growth medium, and only small amounts of formic acid were produced as an end product. Without selenium, formic dehydrogenase activity was small while high amounts of formic acid were produced. Since growth was not affected, formic acid synthesis is regarded as independent of formic dehydrogenase. Formic acid in *E. coli* is formed from pyruvate via pyruvate formatelyase, which is normally absent in *Clostridium,* and oxidized to CO_2 and H_2 via the formic hydrogenlyase system. Formic acid accumulates in the medium in the early growth phase indicating that the formic hydrogenlyase system is only formed in the latter growth phase. This was confirmed by the finding that hydrogen formation and the decrease in formate concentration began simultaneously in the middle of the growth phase. Recent evidence [107] indicates that formic dehydrogenase in *E. coli* is a membrane-bound molybdoprotein. Treatment of the membrane with deoxycholate released the formic dehydrogenase in soluble form complexed with cytochrome b_1.

c) Hydrogenase

The enzyme hydrogenase is present in many microbes. Stephenson and Stickland (1931) [390] reported its presence in *E. coli* and it has been detected in various animal [238, 225], plant [339], and bacterial cells [264]. Activity is measured by the ability to reduce methylene blue with H_2 or by oxidizing chemically reduced dyes, in particular methyl viologen, or the reverse reaction can be measured by following the evolution of H_2 [115]. Rumen species of the following genera produced hydrogen gas from dithionite-reduced methyl viologen: *Bacteroides clostridiiformis; Butyrivibrio fibrisolvens, Eubacterium limosum, Fusobacterium necrophorum, Megasphaera elsdenii, Ruminococcus albus, Ruminococcus flavefaciens, Clostridium pasteurianum, Escherichia coli.* Only *C. pasteurianum, B. clostridiiformis, E. limosum,* and *M. elsdenii* produced hydrogen gas from dithionite [182]. In addition, all of the above species except *E. coli* produced hydrogen gas from pyruvate. Hydrogenase catalyzed reactions are reversible in most cases [135, 123, 298].

$$H_2 \leftrightarrows 2\,H^+ + 2\,e^- . \qquad (17)$$

Those microorganisms demonstrated to utilize hydrogen presumably have the potential to evolve hydrogen as well. Packer and Vishniac (1955) [293] first established NAD as the specific electron acceptor of the purified hydrogenase of *Hydrogenomonas.* It was found possible to produce hydrogen gas from NADH and that hydrogenase mediated the reversible reaction between hydrogen and NAD (Fig. 7). A few reductants have been shown to be immediate electron donors for hydrogenase. These include reduced ferredoxin [442, 273]; cytochrome c_3 [482]; and reduced methyl viologen [482, 273].

Fig. 7. The mediation of the reversible reaction between hydrogen and NAD by the enzyme hydrogenase (Packer and Vishniac, 1955 [293]; Bone, 1963 [34])

Ferredoxin is usually reduced by a coupled reductase reaction making reduced ferredoxin (Fd) the true reductant, and other reactants, the oxidants such as NAD [417], NADP [183], formate or pyruvate [298].

Mortenson and Chen (1974) [264] have recently reviewed the physical and chemical properties of hydrogenase and its role in energy metabolism.

d) Nitrogenase

Biological nitrogen fixation occurs among bacteria belonging to certain *Bacillus, Clostridium* or photosynthetic blue-green algae such as species of *Anabaena* and *Nostoc* [388] when grown under anaerobic conditions. Bacteria of the genus *Rhizobium* also fix nitrogen.

Nitrogen fixation is catalyzed by an enzyme called nitrogenase which converts chemically inert nitrogen gas into combined nitrogen. The nitrogenase reaction has been studied in over sixteen organisms including anaerobes, aerobes and photosynthetic organisms [149] and is shown to be similar in all organisms. Nitrogenase is an enzyme complex consisting of two proteins, a molybdenum-iron protein and an iron protein. The pathways of nitrogen fixation have been reviewed by Benemann and Valentine (1972) [24]. A source of low potential electrons and ATP is required in order for nitrogenase to reduce N_2 to ammonia. The electrons are supplied by reduced ferredoxin or flavodoxin [245]. A schematic representation of the nature of the nitrogenase reaction is shown in Fig. 8. Nitrogenase reduces a variety of substrates and not just N_2. There is an absolute requirement of Mg-ATP to effect reduction of the "Nitrogenase substrates". Nitrogenase catalyzes an ATP-dependent H_2 evolution in the absence of N_2 if a source of electrons is present. H_2 evolution by nitrogenase has been termed ATP-dependent H_2 evolution in *Clostridium pasteurianum* [59, 56, 57, 146, 148, 194, 470].

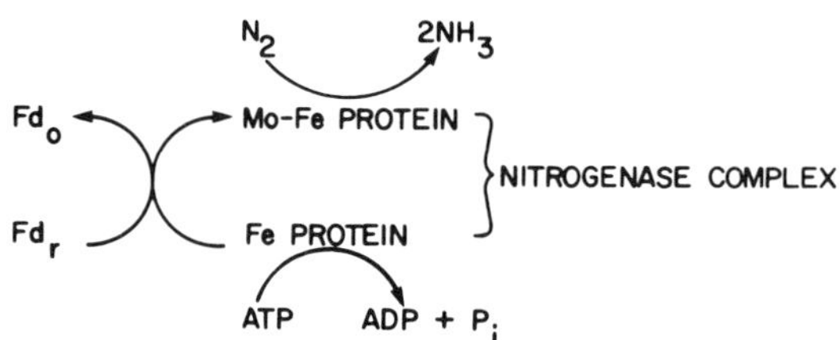

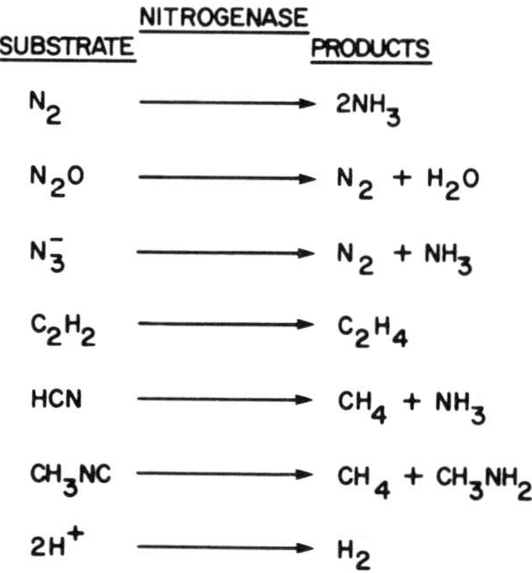

Fig. 8. Schematic representation of the nature of the nitrogenase reaction (Lovenberg, 1974 [245]) and the reactions catalyzed by nitrogenase (Burris and Orme-Johnson, 1974 [60]). Fd = ferredoxin, o = oxidized form, r = reduced form

The specific inhibitors of N_2 fixation can be classified as those which are alternative substrates and those which are not. The first group includes N_2O, NaN_3, C_2H_2, HCN, CH_3CN, and the second group includes H_2, CO, NO and analogs of these compounds [60]. Nitrogenase activity is routinely measured by the ability of the enzyme to reduce acetylene to ethylene [149]. H_2 is a competitive inhibitor of nitrogen fixation [454, 60] and N_2 inhibits H_2 evolution (reciprocal inhibition). Hydrogen evolution is catalyzed by CO sensitive ATP-independent hydrogenase and by nitrogenase in a CO-insensitive ATP-dependent reaction in *C. pasteurianum* (Fig. 9). Hydrogen gas has no effect on the use

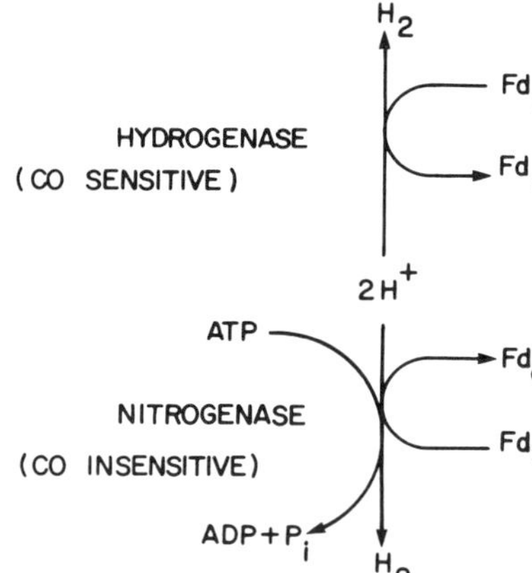

Fig. 9. A schematic diagram of hydrogen evolution by carbon monoxide (CO)-sensitive, ATP-independent hydrogenase and by a CO-insensitive, ATP-dependent nitrogenase reaction in *Clostridium pasteurianum*

of fixed nitrogen compounds by N_2-fixing organisms. However, in the presence of H_2, hydrogenase can reduce ferredoxin and support N_2-fixation. The N_2-fixing system of *Clostridium pasteurianum* is regulated by NH_4^+ [89]. No detectable nitrogenase exists in this organism when grown in the presence of excess NH_4^+. The addition of NH_4^+ to a N_2-fixing culture causes an abrupt halt to nitrogenase biosynthesis. Similar findings have been reported [402] for *Azotobacter vinelandii.* NH_4^+ however, does not affect the *in vitro* nitrogenase activity. Although carbamyl phosphate inhibits nitrogenase activity and its synthesis [262], no effect on H_2 evolution catalyzed by nitrogenase occurred [362, 363].

4. Electron Carriers

By 1965 there was ample data to establish that the formic hydrogenlyase enzyme complex of *coli-aerogenes* group of bacteria was comprised of at least two enzymes, a soluble formate dehydrogenase and particulate hydrogenase, and two unidentified intermediary electron carriers designated as cytochrome reductase and cytochrome c [133] as shown in Fig. 10. The concomitant appearance of cytochrome c_3, hydrogenase and hydrogenlyase activity with added hemin to anaerobic cultures of heme *E. coli*

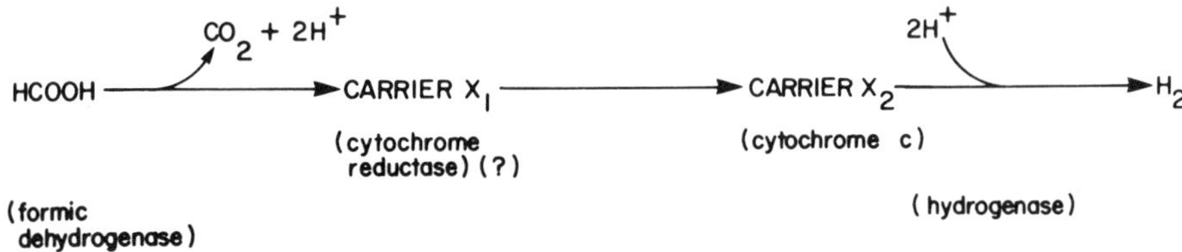

Fig. 10. The anaerobic breakdown of formic acid by *Escherichia coli* with the involvement of two unidentified intermediary electron carriers, cytochrome reductase and cytochrome c (Gray and Gest, 1965 [133])

mutants", suggested a pivotal role for cytochrome c in the production of H_2 [132]. Further work [466, 78, 468] confirmed cytochrome c_{552} in *E. coli* as one of the electron carriers involved in the formic hydrogenlyase system. Other electron carriers were discovered as possible candidates and were believed to act as coupling factors between formic dehydrogenase and hydrogenase, flavodoxin and ferredoxin in *E. coli* [445, 209]. Figure 11 shows the E_0 values for the half-reactions involving these and other electron carriers.

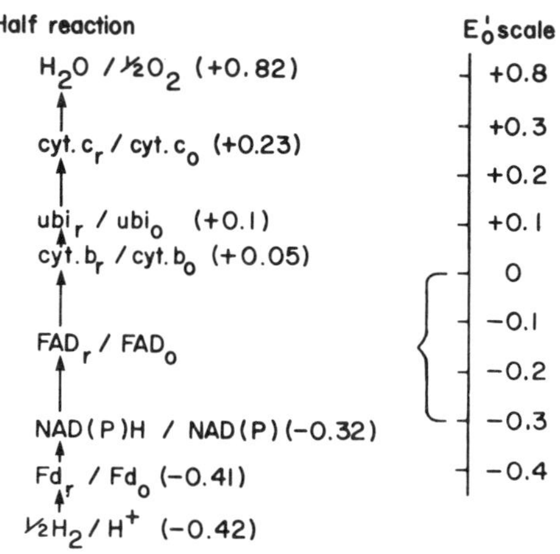

Fig. 11. The E'_0 values for various half-reactions involved in substrate dehydrogenation, e.g. H_2, Fd_r, etc., and electron transport. Fd = ferredoxin, NAD(P) = nicotinamide adenine dinucleotide (phosphate), FAD = flavine adenine dinucleotide, cyt.b = cytochrome b, cyt.c = cytochrome c, r = reduced form, o = oxidized form

The production of H_2 from glucose by *Clostridium butyricum* was found to be inhibited by carbon monoxide [123]. Reversibility of the inhibition by light was considered to be typical of an iron protein [449]. In the presence of CO, the degradation of glucose resulted mainly in lactic acid production and not in the production of H_2, CO_2, and volatile acids. Cells cultivated in a medium deficient in iron also carried out a lactic acid fermentation [14]. Thus an iron protein was indicated as an additional electron carrier required for H_2 production from pyruvate [473, 298, 266]. Investigations indicated that the iron-sulfur protein ferredoxin mediated the electron flow from the thioclastic cleavage of pyruvate to hydrogenase in *Clostridium pasteurianum* [259, 260, 440, 442].

Hydrogenase catalyzes the oxidation of reduced ferredoxin and the reduction of protons producing molecular hydrogen. The protons act as terminal electron acceptors. Confirmation of a direct linkage between ferredoxin and hydrogenase was obtained in studies by Chen and Mortenson (1974) [71]. Flavodoxin [208], NAD [261, 264] and an artificial redox dye have been found able to replace or support ferredoxin in H_2 evolution, pyruvate metabolism and N_2-fixation in *Clostridium pasteurianum.* Excess electrons produced in the form of NADH can be transferred via a flavo-protein reductase from NADH to ferredoxin. Reduced ferredoxin is then oxidized by hydrogenase with the formation of hydrogen gas (Fig. 12).

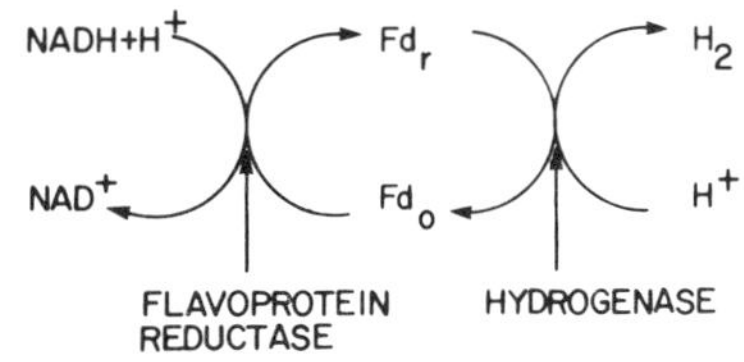

NADH+H$^+$ → Fd$_r$ → H_2
NAD$^+$ ← Fd$_o$ ← H^+
FLAVOPROTEIN REDUCTASE
HYDROGENASE

Fig. 12. The support of ferredoxin in H_2 evolution with excess electrons in the form of NADH (Mortenson, 1968 [261]; Mortenson and Chen, 1974 [264])

Anaerobic and strongly reducing conditions proved a suitable environment of a low redox potential (–420 mV) for ferredoxin. Preparation of clostridial hydrogenase can also reduce pyridine nucleotides with H_2 by a reversible ferredoxin-dependent process [440, 442]. The importance of ferredoxin in H_2 production has been further demonstrated in studies with ferredoxin-depleted preparations of *Bacillus polymyxa* [365]. The addition of ferredoxin from *B. polymyxa* restored the ability to evolve H_2. The strongly reducing conditions for this system can also shunt the electron flow toward the reduction of pyridine nucleotides for biosynthetic reactions or the reduction of nitrogen to ammonia in nitrogen-fixing organisms or the reduction of protons to hydrogen gas [245]. As a result of these and other discoveries, ferredoxin has a central role in the transfer of electrons in oxidation-reduction reactions of anaerobic metabolism (Fig. 13).

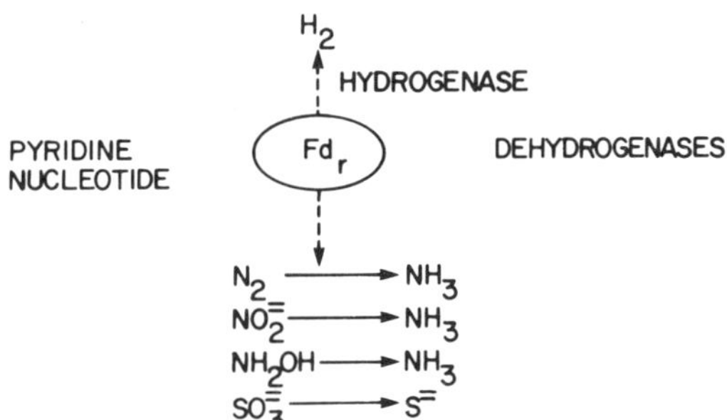

Fig. 13. Central role of ferredoxin (fd) in anaerobic metabolism (after Lovenberg, 1974 [245])

Efforts have been made to determine if electron carriers other than non-heme compounds are present in certain anaerobes. Cytochrome b has been recently demonstrated in homoacetate fermenting species of *Clostridium,* which is the first report of the presence of cytochromes in *Clostridium* [130]. To what extent these heme proteins exist in other clostridia and their role in H_2 production has not yet been determined.

5. Inhibitors of Hydrogen Gas Production

The production of H_2 gas in both clostridial and *coli*-type bacteria is a result of the cell disposing of excess electrons released during pyruvate oxidation. The results of experiments with various factors inhibiting molecular hydrogen production published over the years have led to strongly support this conclusion. There are basic similarities and differences between the clostridial and coli mechanisms of H_2 production, and hydrogen gas inhibition of H_2 production. Hydrogen gas [224, 213, 241, 258], oxygen gas [488, 187], carbon monoxide [123] and media deficient in iron [450, 14] can all effectively inhibit H_2-gas production in *E. coli* and clostridia.

A partial pressure of 25% hydrogen in the atmosphere changes the entire course of the *Clostridium* fermentation. With *C. butylicum* the result is that lactate is formed in place of acetate, butyrate, CO_2 and H_2 [70]. In addition hydrogen gas inhibits the growth of the hydrogen-producing culture, *C. cellobioparum* but not in *E. coli* [72]. The inhibition of H_2 production was found reversible upon the removal of hydrogen gas. The reversibility of H_2 production in *C. butylicum* was indicated by incorporation of isotopic-CO_2 into the carboxyl group of pyruvate [465] and by the inhibiting effect of high H_2 pressure on the rate of the forward reaction of pyruvate decomposition [224, 258]. The presence of carbon monoxide altered the degradation of glucose by *C. butyricum.* Lactic acid was produced rather than volatile acids, CO_2 and H_2 by the inhibited cells. The same result was found [14] by controlling the iron nutrition of this microorganism, and in *E. coli* [450]. In the latter study formate accumulated in the medium. Iron deficient cells demonstrated little or no hydrogenase activity, and no conversion of formic acid to H_2 and CO_2.

6. Microorganisms Involved in Hydrogen Synthesis

a) Clostridium kluyveri

Bornstein and Barker (1948) [36, 37] examined a special type of hydrogen producer. It was the *Clostridium kluyveri* isolated from a sample of black mud at 30 °C. Sulfate reducing and methane producing bacteria were eliminated as far as possible by replacing most of the sulfate in the medium by chloride and reducing the level of CO_2 (carbonate) to 1 mg/100 ml. *C. kluyveri* demonstrated requirements for CO_2 and for two simple organic compounds for its energy metabolism. The two-substrate requirements results from the need for one compound (ethanol) as a reductant and another (acetate) as an oxidant. Acetate could have been replaced by propionate and, less adequately by butyrate. *C. kluyveri* did not attack glucose, lactate or pyruvate in a synthetic medium supplemented with biotin and p-amino benzoic acid. The CO_2 consumed during growth was found to be used for the synthesis of cellular constituents and not for fatty acids. The microorganism catabolized acetate producing butyrate, caproate and small amounts of hydrogen gas [380]. Vinylacetate can also be oxidized anaerobically with the formation of H_2 [307]. Gest (1954) [123] indicates that the precursor of H_2 is a C_2 compound such as an acetaldehyde derivative (Fig. 14). An aldehyde dehydrogenase was found to catalyze the oxidation of acetaldehyde in *C. kluyveri* which is activated by either NAD^+ or $NADP^+$ [61, 383, 156]. Aldehyde dehydrogenase catalyzes the reversible dehydrogenation of acetaldehyde to acetyl CoA [61, 383]. In other bacteria aldehyde dehydrogenase is responsible for the formation of ethanol from acetyl-CoA [90]. In the earlier studies [62, 383], acetyl-CoA was not found to support ATP generation. This view is currently not supported. Thauer *et al.* (1968) [416] found that ATP was generated

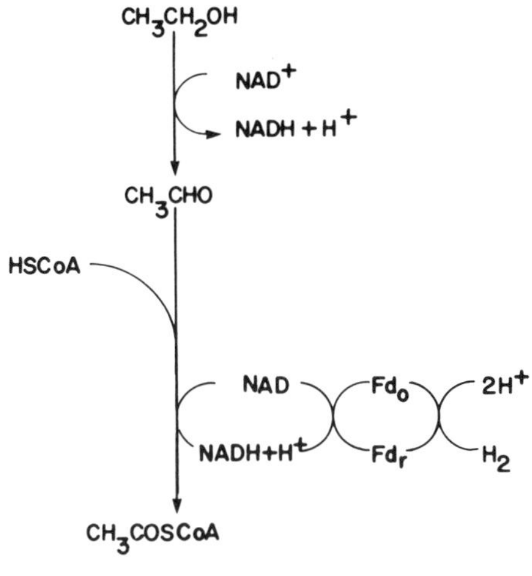

CH_3CH_2OH
NAD^+
$NADH + H^+$
CH_3CHO
HSCoA
NAD
Fd_o
$2H^+$
$NADH+H^+$
Fd_r
H_2
$CH_3COSCoA$

Fig. 14. The pathway of ethanol oxidation to acetyl-CoA by *Clostridium kluyveri*

exclusively from acetyl-CoA in *C. kluyveri* and that two mols of H_2 were evolved per mol of ATP generated. Electron transport phosphorylation is not involved with ATP synthesis in *C. kluyveri.* In cell-free systems of *C. kluyveri,* Jungerman *et al.* (1969) [183] described the ferredoxin-mediated formation of H_2 from NADPH. Acetyl-CoA in addition to ferredoxin were required for H_2 evolution from NADH; pyruvate supported CoA and ferredoxin-dependent H_2 evolution; and dithionite supported CoA and NADH independent H_2 evolution [417]. The enzyme pyruvate ferredoxin oxidoreductase catalyzes the formation of pyruvate from acetyl-CoA and CO_2 in *C. kluyveri* [418, 419] rather than the reverse reaction found in Clostridium.

Therefore, ferredoxin-mediated H_2 evolution from reduced pyridine nucleotides is coupled to ATP formation. Control of ferredoxin reduction influences the rate of energy transformation.

Studies indicate that hydrogen evolution is regulated by the acetyl-CoA/CoA ratio in *C. kluyveri* [417, 185]. Since acetyl-CoA ist an allostearic activator of NADH : ferredoxin oxidoreductase, the acetyl-CoA/CoA level regulates the activity of NADH : ferredoxin oxidoreductase which in turn regulates the NADH/NAD system by consumption of NADH, and the movement of acetyl-CoA towards ATP formation, reduction to fatty acids and hydrogen gas evolution [425, 183].

The formation of stoichiometric amounts of acetate and H_2 must be coupled to phosphorylation provided the H_2 formed is continuously removed to maintain a low H_2 partial pressure and make the reaction (acetaldehyde oxidation) thermodynamically feasible [425].

The ethanol-acetate uptake of *C. kluyveri* was analyzed with respect to possible ATP-yielding reactions and to the significance of the evolution of hydrogen gas [356]. The following conclusions were presented: (a) hydrogen gas is an essential end product of the oxidation of ethanol-acetate, (b) one mol of acetyl-CoA becomes available to the cells for ATP synthesis for every two mols of hydrogen gas evolved, (c) hydrogen gas is formed in the dehydrogenation of acetaldehyde, (d) less than one mol of hydrogen gas is formed per mol of acetaldehyde oxidized indicating that acetate is required for the oxidation of ethanol.

b) Methanobacterium sp.

Bryant *et al.* (1967) [50] presented evidence that *Methanobacterium omelianskii* supplied by Barker (1956) [19] was actually a mixture of two distinct species. One distinct organism (S organism) oxidizes ethanol to acetate with the formation of H_2,

$$CH_3CH_2OH + H_2O \rightarrow CH_3COOH + 2\ H_2 \tag{18}$$

and a methanogenic organism oxidizes H_2 and reduced CO_2 to CH_4 (strain MOH). An atmosphere of hydrogen gas (0.5 atm H_2) inhibited growth of *M. omelianskii* on pyruvate. Cytochromes were not detected in the S-organism. *M. omelianskii* S-organism grew well on pyruvate, oxaloacetate and acetaldehyde. Sulfate, nitrate or other electron acceptors could not replace protons (H^+) as alternate electron sinks during growth on ethanol. Cell-free extracts of S-organism contain alcohol dehydrogenase and catalyze a ferredoxin-dependent oxidation of NADH to NAD and H_2 [333]. An NAD-linked formic dehydrogenase was also discovered with ferredoxin-dependent hydrogenase oxidation of NADH to NAD and H_2. Ferredoxin was essential for the H_2 production aspect of this formic hydrogenlyase system, but not for CO_2 production, and not for NAD reduction. The scheme for electron transfer from formate to H_2 in the S-organism is described in Fig. 15a. Reddy *et al.* (1972) [333] pointed out the similarities between the pyridine nucleotide-linked hydrogenase system of the S-organism to that present in *Clostridium kluyveri* [113, 202]. Further study showed the conversion of acetaldehyde to acetate and H_2 via aldehyde : ferredoxin oxidoreductase activity and by the ferredoxin-linked hydrogenase system. A scheme for transfer of electrons from ethanol and acetaldehyde to H_2 [334] in *Methanobacterium omelianskii* S-organism is shown in Fig. 15b. The

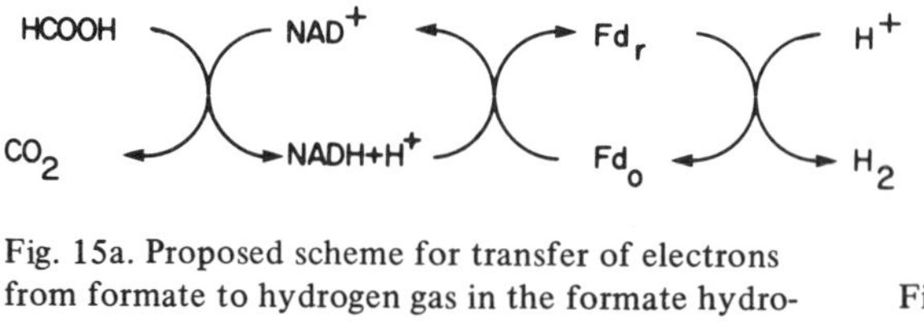

Fig. 15a. Proposed scheme for transfer of electrons from formate to hydrogen gas in the formate hydrogenlyase system of S-organism (Reddy *et al.*, 1972 [333]). Fd = ferredoxin, r = reduced form, o = oxidized form

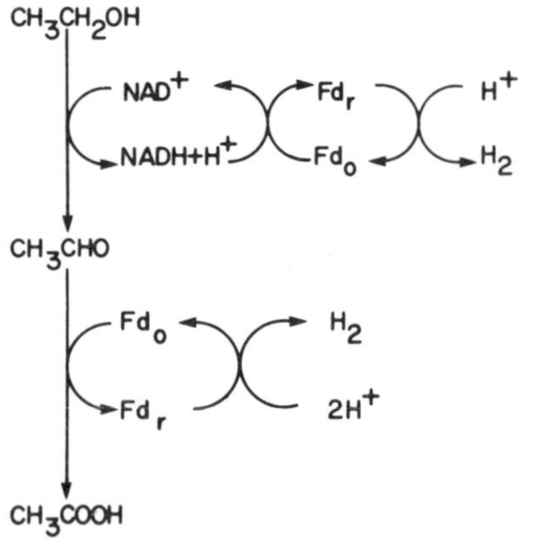

Fig. 15b. Proposed scheme for ethanol metabolism by the S-organism (Reddy *et al.*, 1972 [334]). Fd = ferredoxin, r = reduced form, o = oxidized form

oxidation of acetaldehyde in *C. kluyveri* is dependent upon NAD, but in the S-organism it appears to be dependent upon ferredoxin rather than NAD, CoA or inorganic phosphate [46, 333, 334].

Most known methanogenic bacteria that use hydrogen as an energy source have been shown to contain factor F_{420} which has a molecular weight of about 620 and exhibits a strong UV-adsorption maximum at 420 nm. Tzeng *et al.* (1975b) [432] demonstrated that *M. ruminantium* possessed a formate dehydrogenase linked to F_{420} as the first anionic electron transfer coenzyme. Reduced F_{420} obtained from the formate dehydrogenase can be linked to the formation of hydrogen by a F_{420}-dependent hydrogenase reaction. This apparently constitutes a simple formate hydrogenlyase system. An NADP-linked F_{420}-dependent hydrogenase system also exists in *M. ruminantium* [431]. Cell-free extracts metabolized formate with the formation of approximately stoichiometric amounts of H_2 and CO_2 [432]. The transport of electrons between formate, hydrogen and NADP (F_{420}: oxidoreductase) in *M. ruminantium* involve pathways which require F_{420} as a necessary intermediate carrier. The reaction mechanisms involving hydrogenase activity seem to be essentially the same whether hydrogen evolution or hydrogen uptake systems are being studied.

c) Veillonella

The genus *Veillonella* is characterized by being anaerobic, gram-negative, and coccoid. It is abundant in the saliva and intestines of man and animals. *Veillonella alcalescens* is further characterized [344–346] by the fact that glucose and other carbohydrates are not fermented, nitrates are reduced and H_2S is produced. Propionic and acetic acids, CO_2 and H_2 are produced from lactate [109, 178, 344]. *Veillonella alcalescens* was found to possess a nitrate reductase system which has characteristics of both assimilatory and respiratory nitrate reduction [174]. Pyruvate and nitrate gave a better growth rate than cells grown on pyruvate alone. Growth can occur in a medium containing nitrate as the source of nitrogen and hydrogen gas as the source of energy. Cells grown on hydrogen and nitrate required 1% w/v casein amino acids for growth, no growth occurred in the absence of nitrate. Ferredoxin is involved in a pyruvate synthase (pyruvate: ferredoxin oxidoreductase) reaction which catalyzes the reversible production of acetyl-CoA and CO_2 from pyruvate [460, 443]. H_2 production from pyruvate and α-keto-glutarate involves the hydrogenase-ferredoxin system. Hydrogen gas has been reported in the conversion of hypoxanthine to xanthine by xanthine oxidase [457]. Ferredoxin and hydrogenase may have been present in the test system. The decomposition of hypoxanthine to xanthine is inhibited in an atmosphere of H_2 [457].

d) Selenomonas

Selenomonas ruminantium is a non-spore forming anaerobe that ferments carbohydrates primarily to lactate, propionate, acetate and CO_2. Hydrogen production has been reported [353] in trace amounts in some strains. Increased amounts of H_2 gas evolved when the selenomonads are co-cultured with methane-producing *Methanobacillus omelianskii* MOH. The methane bacteria reduce CO_2 with H_2 to CH_4.

Free H_2 did not accumulate in the combined cultures and the increased amount of H_2 formed by *S. ruminantium* was reflected in the amount of CH_4 produced by the methanogenic bacteria. The presence of a nicotinamide adenine dinucleotide-linked hydrogenase activity suggested the H_2 is a product of the oxidation of NADH, and that the formation of H_2 from NADH is increased when H_2 is removed by growth with a methanogenic culture.

e) Ruminococcus albus

The major products of digestion by the mixed microbial population in primary sewage sludge or in the rumen are acetic propionic, and butyric acids in addition to H_2, CO_2, and CH_4 [169, 409]. Hungate (1966) [169] hypothesized that electrons generated by the oxidation of the substrates by one group of bacteria in the rumen can be somehow made available for use by another group of H_2-utilizing bacteria also present in the rumen. Iannotti *et al.* (1973) [173] tested this hypothesis with a mixed culture of *Ruminococcus albus* and *Vibrio succinogenes*. *R. albus* is a cellulolytic bacterium that produces ethanol, acetate, formate, H_2 and CO_2 from cellobiose [49, 173]. *R. albus* does not produce H_2 from formate, but has been shown to produce formate from CO_2 and H_2 rather than from pyruvate [255, 182]. *V. succinogenes* cannot utilize glucose, ethanol or acetate as energy sources in the presence or absence of fumarate [475], but it can couple the oxidation of H_2 or formate with the reduction of fumarate to succinate (Iannotti *et al.* (1973) [173]). *R. albus* does not reduce fumarate. As a chemostat co-culture, *R. albus* and *V. succinogenes* produced succinate, no ethanol and increased amounts of acetate accumulated. As a result the quantity of succinate produced was considered to reflect the quantity of H_2 evolved by *R. albus* representing 4 mols H_2 compared to 2.6 mols H_2 produced by *R. albus* in pure culture. In addition, for every shift from ethanol to acetate formed in the mixed culture, an extra ATP is generated via acetyl-CoA [173, 425].

f) Trichomonas and Eukaryotic Organisms

Trichomonads are symbiotic or parasitic flagellates that live in the digestive or in the genitourinary system of the host. Their metabolism is essentially anaerobic, but they tolerate the presence of oxygen. Protozoa adapted to anaerobic conditions contain no morphological recognizable mitochondria but do form microbody-like structures which represent the redox organelles in trichomonads [267]. These microbodies are called hydrogenosomes based on their biochemical composition [238].

The anaerobic catabolism of pyruvate by *Tritrichomonas foetus* resembles that observed in some clostridia. The products are acetate, succinate, CO_2 and H_2. *T. foetus* apparently lacks the means to produce formate from pyruvate or to utilize formate in the production of CO_2 and H_2, as evidenced by the lack of pyruvate: formate lyase and formic dehydrogenase [238]. Low redox potential-reducing equivalents are generated in the decarboxylation of pyruvate and are transferred to an unknown acceptor, possibly ferredoxin or flavodoxin. The reduced acceptor is reoxidized by a hydrogenase and molecular hydrogen is formed.

g) Photosynthetic Bacteria

The oxidation of water in plant photosynthesis results in the production of oxygen and the transfer of electrons. This oxidation does not occur in photosynthetic bacteria. The bacterial photochemical reaction apparatus can mediate cyclic photophosphorylation and the reduction of pyridine nucleotide provided that light and suitable electron donors are available, such as H_2S, H_2, or organic compound. The bacterial photochemical reaction apparatus consists of a photosynthetic electron transport chain as shown in Fig. 16. The photosynthetic electron transport chain includes ferredoxin, flavoproteins, quinones, and cytochromes. The energy conversion and generation of reducing

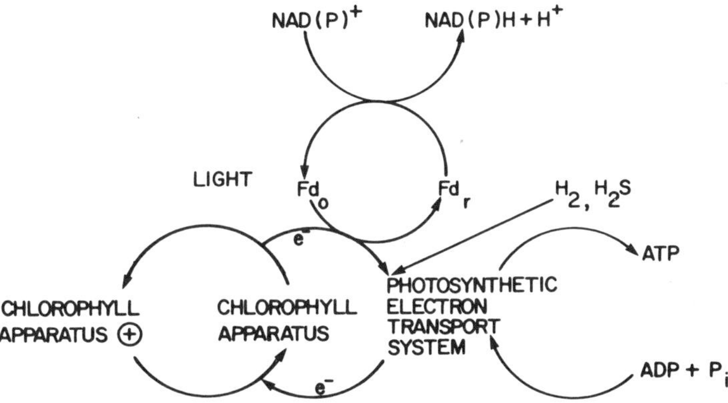

Fig. 16. A schematic diagram showing the mechanism of photophosphorylation and pyridine nucleotide and ferredoxin reduction. Fd. = ferredoxin, o = oxidized form, r = reduced form

power in bacterial photosynthesis is discussed by Gest (1972) [126]. Ferredoxin was found in *R. rubrum* as early as 1962 [411]. Its role in carbon assimilation was investigated by Buchanan *et al.* (1967) [54].

Sulfur and non-sulfur bacteria evolve H_2 in the dark while using glucose, C_3 compounds or formate [269]. H_2 and CO_2 were produced when growth by the non-sulfur purple bacterium *Rhodospirillum rubrum* occurred with glutamate or aspartate instead of ammonium as nitrogen source [119, 120]. The growth medium contained dicarboxylic acids of the citric acid cycle. Since that time extensive reviews on the photoproduction of molecular hydrogen by sulfur, non-sulfur purple or green sulfur bacteria have been published [123, 244, 387, 133, 287, 308, 264]. These microorganisms all contain nitrogenase, the enzyme that can also reduce protons to form H_2 [60, 264]. The evolution of H_2 occurs when N_2 is absent or ammonium becomes limiting, and when ATP from photophosphorylation and reductants from acetate, malate, succinate or fumarate oxidation are in excess [286, 125, 39, 308, 261].

Photoevolution of H_2 is inhibited by N_2 and NH_4^+ [270, 119, 41, 286, 156, 157] or when high concentrations of yeast extract or peptone are present [123]. The cessation of H_2 evolution upon the addition of ammonia was considered to be the result of the reductive amination of organic compounds at the expense of reductants such as NADH [261]. When N_2 is present most of the available electrons are used in N_2 reduction at the expense of H_2 formation. A high ratio of reduced to oxidized NAD was considered of little importance for H_2 production since cultures of *R. rubrum* utilizing fructose have a very high level of NADH and only little H_2 formation [357]. Ammonium salts provided as a nitrogen source repress synthesis of the hydrogen-evolving system [286]. When the N/C ratio in the growth substrates (i.e. glutamate) exceeds a critical value, free ammonia appears in the medium and H_2 is not evolved [157]. The reduction of protons by nitrogenase and the synthesis of further nitrogenase ceases [276, 279, 157, 158].

H_2 is a competitive inhibitor of nitrogenase and nitrogen fixation [454], but H_2 does not inhibit nitrogenase catalyzed H_2-evolution [59]. Hydrogenase has been clearly demonstrated in photosynthetic bacteria by the oxidation of H_2 in which ferricyanide acts as an electron acceptor [38]. Resting cells of *R. rubrum* are capable of completely photometabolizing acetate, fumarate, malate and succinate to CO_2 and H_2 [287]. The photoassimilation of hydrogen and carbon dioxide in *Rhodomicrobium vannielii* was not inhibited by nitrogen gas or ammonium salts [160] which is also indicative of an active hydrogenase.

Cultures of *Chloropseudomonas, Chromatium* and *Rhodospirillum* were described as evolving hydrogen gas using various substrates with L-arginine and glutamate as the source of nitrogen respectively [28, 216]. A thioclastic or formic hydrogenlyase system was also suggested. Hydrogen evolution in *R. rubrum* also occurs simultaneously with nitrogen fixation only under photoheterotrophic conditions [354]. The photoevolution of hydrogen gas was found to be substrate dependent [354]. The rate of hydrogen evolution for one and the same substrate was found to be dependent upon cultural conditions, cultural age and the pretreatment of cells. Under photoautotrophic conditions hydrogen evolution ceased and both CO_2 and N_2 were photoreduced.

The production of small amounts of hydrogen was observed [433, 358, 359a] in *R. rubrum* using anaerobic conditions in the dark. Pyruvate was used as the substrate and was oxidatively decarboxylated. Major end products were acetate, propionate, CO_2 and H_2. Formate was formed as an intermediary product in a strain of *R. rubrum* capable of producing H_2 and CO_2 [359, 359a]. Pyruvate : formate lyase appears to be a characteristic key enzyme of the dark anaerobic fermentation of pyruvate metabolism in *R. rubrum* [359a]. Some purple sulfur bacteria can evolve hydrogen gas in the light as a result of thiosulfate oxidation [126]. A tentative scheme of substrate decomposition and electron transfer in cells of *R. rubrum* under anaerobic conditions in the dark is shown in Fig. 17. *R. rubrum* produces two types of ferredoxin [364]. One (Fd-I) is formed only when the cells are grown photosynthetically, and the other (Fd-II) is formed when cells are grown either photosynthetically or heterotrophically in the dark. Fd-I was found to be greatly superior to Fd-II in promoting the activity of nitrogenase isolated from photosynthetically grown cells of *R. rubrum.* Synthesis of nitrogenase was accompanied by a light-dependent mechanism of H_2 evolution.

The highest yields and rates of formation of H_2 with *Rhodopseudomonas capsulata* were obtained with the organic acids lactate, pyruvate, malate, and succinate in media containing glutamate as the nitrogen source. Under optimal conditions with excess

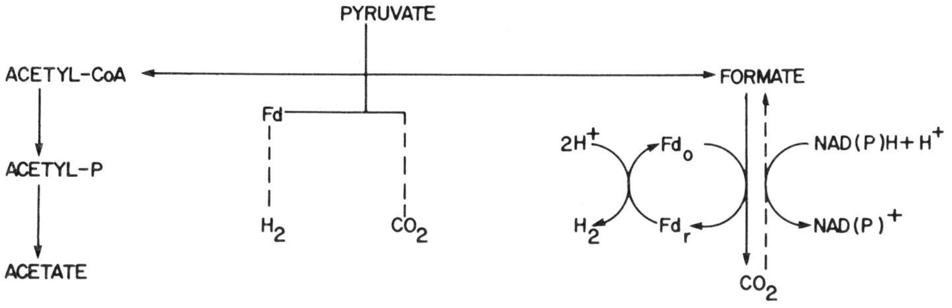

Fig. 17. A tentative scheme of pyruvate decomposition and electron transfer in cells of *Rhodospirillum rubrum* under anaerobic conditions in the dark (Schon and Bidermann, 1973 [359])

lactate, H_2 was produced at rates of 130 ml $\cdot$ g^{-1} dry weight cells $\cdot$ h^{-1} [157, 158]. *R. capsulata* photoheterotrophically evolves H_2 catalyzed by nitrogenase and utilizes H_2 catalyzed by hydrogenase as a reductant for photoautotrophic growth [447]. Formation of H_2 from organic compounds such as lactate, pyruvate, malate, succinate is mediated by nitrogenase and is not inhibited by an atmosphere of 99% H_2 but is inhibited by ammonium [158]. *R. capsulata* and related bacteria employ controls which ensure that when readily utilizable organic H-donors are supplied, the system that catalyzes light-dependent reduction of CO_2 with H_2 (photoreduction) becomes inoperative [158]. The H_2 evolving function of nitrogenase appears to provide a means of coping with excessive fluxes of ATP and reducing power especially under photoheterotrophic conditions when the energy supply is not limiting.

h) Algae

Under natural conditions the absorption of solar radiation by green plants results in the oxidation of water. The oxygen is released in molecular form and the reducing equivalents, in the form of NADPH, are used for the assimilation of carbon dioxide into cell constituents. It is possible under certain conditions for the photosynthetic apparatus to provide electrons released from water at a reducing potential equal to or more negative than the hydrogen electrode. The reducing potential of these electrons can then be coupled with hydrogen ions to form hydrogen gas catalyzed by hydrogenase. A tentative scheme for electron transport in plant photosynthesis is shown in Fig. 18. The electrons from the weak reductant in system-II enter at the level of cytochrome b and as they transverse the path leading to the photoreaction centre of system-I, advantage is taken of the opportunity to generate ATP, before the electrons are further energized in order to reduce NADP [352, 330]. Gaffron and Rubin (1942) [118] found that the hydrogenase containing unicellular alga *Scenedesmus* was capable of reducing carbon dioxide with molecular hydrogen in the light and liberated hydrogen gas in the dark when air was replaced with nitrogen. Illumination of the fermenting algae enhanced the liberation of hydrogen gas, particularly if CO_2 and H_2 were absent. The return of the photosynthesizing cell to aerobic conditions resulted in the cessation of H_2 evolution. The substrate utilized was unknown. *Chlamydomonas moewousii*

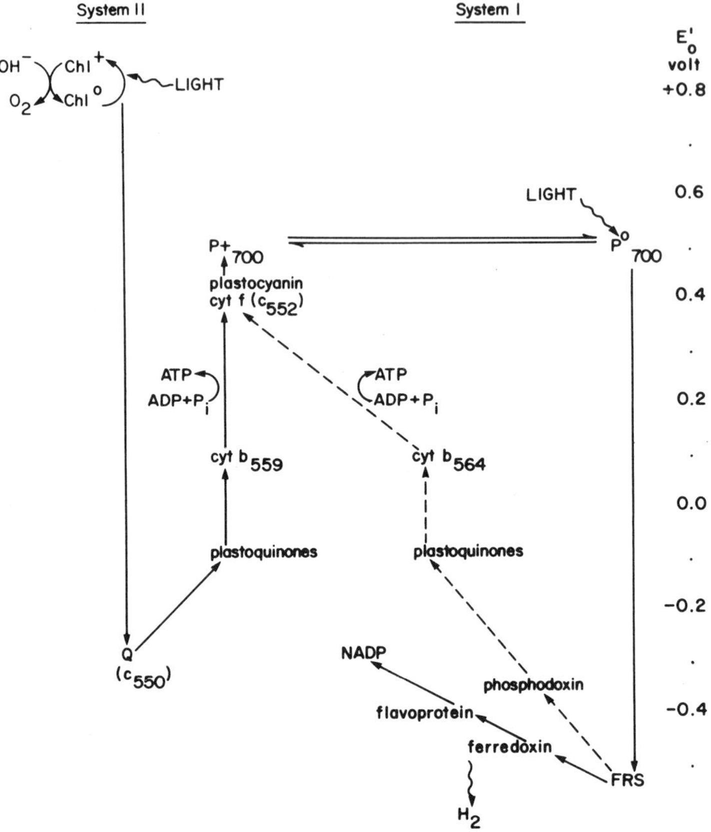

Fig. 18. Tentative scheme for electron transport in plant photosynthesis. The arrows represent the direction of electron flow. Broken arrows represent electron path during cyclic photophosphorylation (White *et al.* 1973 [456]). FRS = ferredoxin reducing substance, cytochrome f = c-type cytochrome (cytochrome c_{552}) in contact with a copper protein plastocyanin, p_{700} = pigment established as the photoreaction centre of System-I and contains chlorophyll, phosphodoxin substance of unkown nature, Q: substance of unknown nature although not a cytochrome. E_0' = oxidation reduction scale

evolved hydrogen and carbon dioxide in the light and dark [114]. One atmosphere of hydrogen gas was found to completely inhibit the photoproduction of hydrogen gas. The presence of nitrogen gas and ammonia had no effect on hydrogen gas evolution.
In *Chlorella,* hydrogenase was activated after a relatively anaerobic period and hydrogen gas evolved upon illumination [377]. At a certain critical level of O_2, hydrogen gas evolution stopped. The light dependent evolution of H_2 from reduced NAD was demonstrated by cell-free extracts of anaerobically adapted *Chlamydomonas* [195]. Similar results were obtained with *Ankistrodesmus* [196], *Chlorella* and *Scenedesmus* [152, 153].
Hydrogen gas evolution with *Chlamydomonas, Chlorella* and *Scenedesmus* in the dark is accompanied by the release of only CO_2 and is depressed by dark starvation and is

inhibited by uncouplers of photophosphorylation [152, 153]. The different effects of uncouplers of photophosphorylation and inhibitors of photosynthesis used in various studies led to the conclusion that the photoproduction of hydrogen gas does not require photosystem-II. Studies by Stuart and Kaltwasser, 1970 [405] indicate an electron flow driven directly by light through System-I from reductant produced in oxidative carbon metabolism to a redox potential capable of reducing protons. H_2 production in *Scenedesmus* does not require cyclic photophosphorylation [406]. Instead, the photo-evolution of H_2 is due to non-cyclic electron flow through a cytochrome b_{552} by photo-system-I to hydrogenase where hydrogen gas is released. This mechanism is similar to that proposed [153] for the photoproduction of hydrogen gas by *Chlamydomonas moewousii.* At this point certain key problems were identified: (1) the identity of the substrate(s) used in H_2 evolution, and (2) the exact site of entry of electrons into the electron transfer chain. Oshchepkov and Krasnovskii (1972) [289] investigated O_2 and hydrogen gas evolution during illumination of *Chlorella pyrenoidosa.* Their work confirmed the effects of light intensity, glucose utilization, wavelength, heating, uncouplers and inhibitors on photoproduction of H_2 reported by previous investigations [188, 1, 152, 153]. The data presented are in conjunction with the scheme previously proposed [125, 133] to explain the photoproduction of H_2 in purple bacteria.

Kessler (1973) [199] suggests that H_2 evolution enables the organism to dispose of excessive amounts of reducing power. Photosystem-II under anaerobic conditions remains reduced and inactive. The presence of hydrogenase enables the reduced photo-system-II to become oxidized in a few seconds, whereas in algae without hydrogenase it remains reduced and inactive. Inhibitor studies of photosystem-II with a nitrogen-fixing blue-green algal symbiont demonstrated that acetylene reduction and hydrogen gas evolution are derived from water by utilizing photosystem-II [279]. Photoevolution of H_2 is stimulated by uncouplers demonstrating that the role of light is not to provide energy by phosphorylation. The anaerobic adaptation of algae is not an essential pre-requisite for manifestation of H_2 evolution under conditions of illumination [396]. Such adaptation is desirable for an increase in the yield of H_2 but it can be replaced by the addition of glucose. Lost photosynthetic activity in blue-green alga (*Phormidium*) was restored by the addition of agents such as cystine, thioglycollate and sulfite which lowered the redox potential [453]. Under anaerobic conditions and in the dark the cell changes to a heterotrophic mode of metabolism as a result of the enzymatic oxidation of endogenous reserves of carbohydrates which are the products of prior photosynthetic activity. The products of glucose fermentation by memebers of the genus *Chlorella* are described [196, 197, 198, 446]. Formic acid and H_2 appeared in those species of *Chlorella* which contain hydrogenase.

Efforts have been made [26] to demonstrate photosynthetic hydrogen production as a method of solar energy conversion in *Anabaena cylindrica.* The oxygen generated from water by the photosynthetic apparatus eventually inhibits hydrogen evolution. Both the hydrogen and oxygen are produced photochemically from water [27]. Hydrogen evolution was inhibited rather by the presence of nitrogen, and was insensitive to CO characteristic of the hydrogen evolving reaction catalyzed by nitrogenase. Oxygen-sensitive nitrogenase is protected against oxygen inactivation by some unkown mechanism operating in the heterocysts of *Anabaena.* One of the problems encountered in the cultures of *Anabaena* was that more oxygen than hydrogen is produced. The heterocyst cell itself has no oxygen-producing apparatus, but it does contain the hydrogen generating apparatus. Reductants are made in the green part of the plant which contains the oxygen-producing cells. The reductants diffuse into the non-green heterocyst to evolve hydrogen. Peak hydrogen gas production rates with various green or blue-green algae systems ranged from 4 $\mu l\ H_2 \cdot mg^{-1}$ of dry weight $\cdot\ h^{-1}$ [152, 153] to 32 $\mu l\ H_2 \cdot mg^{-1}$ of dry weight h^{-1} [452]. Periods of peak productivity also ranged from a few hours to a few days. The thermodynamic efficiency of converting incident light energy to free energy of hydrogen via algal photosynthesis was 0.4% [452] with cultures of *Anabaena cylindrica.*

i) Miscellaneous Hydrogen Producing Systems

The major challenge is to use water and sunlight to generate H_2 gas in a biologically-based system utilizing chloroplast membranes and the enzyme hydrogenase as catalysts. Arnon *et al.* (1961) [8] demonstrated the coupling of spinach chloroplasts and partly purified *Chromatium* hydrogenase for photoproduction of H_2, using cysteine as the electron donor.

Benemann (1973) [25] described the evolution of hydrogen from spinach chloroplast preparations mixed with *Clostridium kluyveri* hydrogenase and ferredoxin. H_2O is the source of hydrogen and light the source of energy. About 1.0 mol of H_2 was evolved per hour with 18 μmols of ferredoxin and 480 mg hydrogenase. Problems encountered with the system were the oxygen inactivation of ferredoxin, autooxidation, hydrogenase and instability of the chloroplast preparation. The presence of oxygen was seen as the main reason for inhibition of hydrogen gas evolution.

The evolution of H_2 gas in an in-vitro illuminated chloroplast plus hydrogenase system was shown to function for six and one-half hours at a continuous rate of about 10 μmols H_2/mg chlorophyll per hour [332]. Chloroplasts from various plant species were used. Both *Clostridium* and *E. coli* hydrogenase were used. Heat inactivation and the use of inhibitors of photosystem-II indicated that H_2O was the source of electrons for H_2 gas production [336]. The production of hydrogen gas by biophotolysis of water was studied using chloroplasts and hydrogenases of various kinds under different experimental conditions [338]. Of the hydrogenases tested (*C. pasteurianum, E. coli, Chromatium C, D. dechloropseudomonas*) the clostridial hydrogenase was the most active H_2 producer. H_2 evolution rates at 20 °C of 20 μmols of H_2/mg chlorophyll per hour were achieved, linear up to 3 h. Chloroplasts from spinach, lettuce and tobacco functioned with equal efficiency. Hydrogen gas evolution was found to be inhibited by oxygen, methyl viologen, potassium ferricyanide and is enhanced by bovine serum albumin and ferredoxin.

Bacteriorhodopsin, the only protein present in the purple membrane from *Halobacterium halobium,* acts as a light energy converter [360]. Bacteriorhodopsin was shown to be the biosynthetic precursor of the purple membrane [407]. Under illumination bacteriorhodopsin undergoes a photochemical cycle, accompanied by a proton release and uptake effecting an electrochemical proton gradient across the cell membrane. This proton gradient (pump) drives ATP synthesis and transport processes which can be understood in terms of the chemiosmotic theory of Mitchell (1976) [257]. Photopotential gradients up to 150 mV have been successfully created across the membrane. A potential difference ($\Delta E'$) of approximately 250 mV is required to allow the synthesis of 1 mol of ATP from ADP and P_i [425]. Artificial membranes containing the chromophore of bacteriorhodopsin exhibiting its special spectroscopic properties were made using special techniques [360]. The aim is to construct a purple membrane containing apparatus which performs work at the expense of sunlight. A ΔpH gradient created upon illumination could be utilized to produce H_2 [371]. The proton pumping activity of bacteriorhodopsin can be clearly demonstrated when light-and-oxygen-induced ATP synthesis is blocked by dicyclohexylcarbodiimid (DCCD). Under these conditions bacteriorhodopsin is left working alone and acidification of the medium occurs to a much greater extent than in non-treated cells [360]. The size of the gradient is limited only by passive back diffusion of H^+ ions.

j) Symbionts

Newton (1976) [279] demonstrated that the water present in the fern *Azolla,* in conjunction with a nitrogen-fixing blue-green algal symbiont *(Anabaena azolla),* present in its leaf cavity, was responsible for the synthesis of the hydrogen gas. Plants grown on nitrate evolve hydrogen gas at rates comparable to their total nitrogenase activity (ethylene production).

Hydrogen evolution and nitrogenase were light-dependent and insensitive to carbon monoxide indicating that the major source of hydrogen gas was from activity of algal nitrogenase and not hydrogenase. Nitrogenase was found to be unstable in the presence of oxygen. Growth on nitrate supported nitrogenase activity and hydrogen evolution. The root nodule bacteria *Rhinozobium* forms a symbiotic nitrogen-fixing association with plants such as soybean, clover, alfalfa, and others. The genus *Rhizobium* is characterized by its ability to elicit nodules and fix nitrogen in the roots of legumes. Nodules from plants catalyze H_2-evolution [162, 29], and hydrogenase is found in pea root nodules [97–99]. The nitrogenase is located in cell-free extracts of bacteriods from soybean root nodules and catalyzes H_2-evolution and is ATP-dependent [211]. Keister (1975) [191] found that several strains of *Rhizobium* species and *R. japonicum* reduce acetylene to ethylene which is characteristic of nitrogenase [148, 149]. A low oxygen concentration (0.2 atm) is needed.

7. Anaerobic Digestion

Anaerobic digestion has been utilized throughout the world for the treatment of human and animal manures for the past 100 years [410]. Today, raw and settled municipal sewage or industrial liquid waste are treated by anaerobic digestion [327, 128]. The primary purpose of anaerobic digestion is to lower the BOD (Biological oxygen demand) of the waste [73]. Although the BOD is apparently lowered further with aerobic metabolism, there are several advantages making anaerobic digestion quite attractive [410]. One of these is that much greater BOD loads can be applied [73]. Another advantage is that methane is a normal and major constituent of the gases evolved. For example, researchers at the University of Pennsylvania and United Technologies [166] were able to obtain a methane yield of 2.8×10^6 cal/kg, of dry solid organic waste matter. An overall efficiency of 60% was obtained for the anaerobic digestion process. It is now known that methane is primarily formed as a result of the reduction of CO_2 with H_2 as reductant [19, 385, 474, 415, 496]. The hydrogen gas and carbon dioxide are supplied in the digester as a result of the metabolic activities of non-methanogenic heterotrophic anaerobic microorganisms [409]. But only traces (2% v/v) of hydrogen are usually evolved with continuous flow digesters [104]. However, when methanogenesis is inhibited in anaerobic sewage sludge digesters, H_2 accumulates along with CO_2 [408]. The accumulated quantities of H_2 evolved is never as great as expected, since it is assumed that 4 mols H_2 are required to produce 1 mol of methane.

The explanation for this may be found when the activities of certain H_2-utilizing bacteria are studied. Carbon dioxide and sulfate reducing anaerobic bacteria [53, 425], are H_2-utilizers normally found in anaerobic digesters. They help to maintain H_2-partial pressures lower than standard conditions thereby providing thermodynamically favourable conditions for the continued production of H_2 from various substrates [425, 53, 50] by heterotrophs. In addition, the continued removal of H_2 prevents inhibition of H_2 production due to H_2 accumulation. Pure cultures of *Clostridium thermocellum* degrades cellulose and cellobiose to produce primarily H_2, CO_2, ethanol and acetic acid but cannot utilize glucose [450a]. Approximately 0.85 mols H_2 were produced per mol of anhydroglucose equivalents (monomeric form of cellulose, MW = 162 $g \cdot mol^{-1}$) fermented. Co-cultures of *C. thermocellum* and *M. thermoautotrophicum,* a methane former, produced greater quantities of H_2 reflected in the quantities of methane evolved. The rate limiting step in the degradation of cellulose by *C. thermocellum* was found to be the solubilization of cellulose.

Those food industry waste waters high in BOD must be considered prime candidates for anaerobic digestion and the concomitant production of methane or hydrogen gas. A previous study [409] has indicated that between 20 to 30% (v/v) H_2 is evolved during the early stages of municipal primary sewage sludge digestion. Few studies have been undertaken to evaluate the ability of various industrial liquid wastes to support hydrogen gas production. For example, potato-processing plants that handle millions of pounds of potatoes a day produce waste water with a BOD equivalent to that of a city of 300000 people [164]. In addition, according to Anderson [5], the energy content of the presently collected agricultural wastes, 50% of which is cellulose, would add up to 1.3×10^{14} $kcal \cdot year^{-1}$ at 1.4 $kcal \cdot kg^{-1}$ [7]. Data on the economics and potential energy yields of industrial wastes is lacking, especially in the food industry. In many cases data on the food processing industry waste is included in the agricultural waste category. Prelimininary studies have been made to evaluate the hydrogen gas producing potential of various industry wastes as nutrient substrate for the bacterium *Citrobacter intermedius* [493]. These wastes include brewery waste, corn steep liquor, cheese whey and spent sulfite liquor. *C. intermedius* is a primary sewage sludge isolate that is capable of producing H_2 when grown on various substrates including glucose under anaerobic conditions [491].

This review indicates that hydrogen gas is evolved by many species of bacteria under a wide variety of conditions. While this is so, the optimization of H_2 production for ultimate use in relatively high volume industrial applications has not been studied or accomplished. The following are examples of some industries which could achieve both cost and waste reductions through the utilization of waste materials for hydrogen gas production: 1) grain industries: wastes such as chaff, grain dust, water or rodent spoiled grain could be used as feedstock for a hydrogen gas producing process; 2) meat processing industries: hydrogen production by fermentation of meat byproducts could yield low-cost in house hydrogen for fat hydrogenation or other chemical conversions; 3) vegetable oil and margarine manufactures: hydrogen gas production to carry out hydrogenation; 4) pulp and paper industries: certain cellulolytic bacteria which can produce hydrogen could be used for the conversion of pulp and paper wastes into H_2 with the reduction of COD and BOD values in the final effluents, contributing to environmental acceptability.

References

1. Abeles, F. P.: Plant Physiol. **39**, 164 (1964).
2. Ackrell, B. A. C., Asato, R. N., Mower, H. F.: J. Bacteriol. **92**, 828 (1966).
3. Akagi, J. M.: J. Biol. Chem. **242**, 2478 (1967).
4. Alford, J. S.: Can. J. Microbiol. **22**, 52 (1976).
5. Anderson, C. G.: An introduction to bacteriological chemistry. William Wood and Co., Baltimore, Maryland 1938.
6. Andrews, J. F., Greaf, S. P.: In "Anaerobic Biological Treatment Processes" (Ed. R. F. Gould), Advances in Chemistry Series 105. American Chemical Society, Washington 1971, p. 126.
7. Antal Jr., M. J., Feber, R. C., Tinkle, M. C.: First World Hydrogen Energy Conference Proceedings (Ed. T. Veziroglu), Univ. of Miami, Coral Gables, Florida 1976, p. 3A–69.
8. Arnon, D. I., Losada, M., Nozaki, M., Tagawa, K.: Nature **190**, 601 (1961).
9. Azouly, E., Marty, B.: Euro. J. Biochem. **13**, 168 (1970).
10. Bachofen, R., Buchanan, B. B., Arnon, D. I.: Proc. Nat. 1. Acad. Sci. **51**, 690 (1964).
11. Baldwin, R. L., Chamberlain, G. D., Milligan, L. P.: Fed. Proc. Fed. Amer. Soc. Exp. Biol. **23**, 485 (1964).
12. Baldwin, R. L., Milligan, L. P.: Biochim. Biophas. Acta **92**, 421 (1964).
13. Banks, B. E. C., Vernon, C. A.: J. Theor. Biol. **29**, 301 (1970).
14. Bard, R. C., Gunsalus, I. C.: J. Bacteriol. **59**, 387 (1950).
15. Barker, H. A.: Arch. für Mikrobiol. **7**, 404 (1936).
16. Barker, H. A.: Enzymologia **2**, 175 (1937).
17. Barker, H. A.: J. Bacteriol. **36**, 322 (1938).
18. Barker, H. A.: Ant. van Leeuwen. J. Microbiol. Serol. **12**, 167 (1947).
19. Barker, H. A.: In "Bacterial Fermentations", John Wiley and Sons, New York 1956, p. 245.
20. Barton, L. L., LeGall, J., Peck, H. D.: In "Horizons of bioenergentics" (Ed. A. San Pietro, H. Gest), Academic Press, New York 1972, p. 33.
21. Beauchop, T., Elsden, S. R.: J. Gen. Microbiol. **23**, 457 (1960).
22. Bell, G. R., LeGall, J., Peck, H. D.: J. Bacteriol. **120**, 994 (1974).
23. Benemann, J. R., Valentine, R. C.: Adv. Microb. Physiol. **5**, 135 (1971).
24. Benemann, J. R., Valentine, R. C.: Adv. Microb. Physiol. **8**, 59 (1972).
25. Benemann, J. R.: Fed. Proc. **32**, 632 (1973).
26. Benemann, J. R., Berenson, J. A., Kaplan, J. A., Kamen, M. D.: Proc. Nat' 1. Acad. Sci. **70**, 2317 (1973).
27. Benemann, J. R., Weare, N. M.: Science **184**, 174 (1974).
28. Bennett, R., Rigopouls, N., Fuller, R. C.: Proc. Nat' 1. Acad. Sci. **52**, 762 (1964).

29. Bergersen, F. J.: Aust. J. Biol. **16**, 669 (1963).
30. Biebl, H., Pfennig, N.: Arch. Microbiol. **112**, 115 (1977).
31. Billen, D.: J. Bacteriol. **62**, 763 (1951).
32. Billen, D., Lichstein, H. C.: J. Bacteriol. **61**, 515 (1951).
33. Blackwood, A. C., Neish, A. C., Ledingham, G. A.: J. Bacteriol. **72**, 497 (1956).
34. Bone, D. H.: Biochem. Biophys. Acta **67**, 589 (1963).
35. Boon, W. R.: Outlook on Agricul. **4**, 163 (1964).
36. Bornstein, B. T., Barker, H. A.: J. Bacteriol. **55**, 223 (1948a).
37. Bornstein, B. T., Barker, H. A.: J. Biol. Chem. **172**, 659 (1948b).
38. Bose, S. K., Gest, H., Ormerod, J. G.: J. Biol. Chem. **236**, PC13 (1961).
39. Bose, S. K., Gest, H.: Proc. Nat' 1. Acad. Sci. **49**, 337 (1963).
40. Bregoff, H. M., Kamen, M. D.: J. Bacteriol. **63**, 147 (1952a).
41. Bregoff, H. M., Kamen, M. D.: Arch. Biochem. Biophys. **36**, 202 (1952b).
42. Bresters, T. W., Kruhl, J., Scheepens, P. C., Veeger, C.: FEBS Lett. **22**, 305 (1972).
43. Brewer, C. R., Werkman, C. H.: Enzymologia **8**, 318 (1940).
44. Bridger, W. A.: In "The Enzymes" vol 10 (Ed. P. D. Boyer), Academic Press Inc., New York 1974, p. 581
45. Brill, W. J., Wolin, E. A., Wolfe, R. S.: Science **144**, 297 (1964).
46. Brill, W. J., Wolfe, R. S.: Nature **212**, 253 (1966).
47. Brown, R. W.: Iowa State College J. Sci. **11**, 39 (1936).
48. Brown, T. D. K., Pereira, C. R. S., Stormer, F. C.: J. Bacteriol. **112**, 1106 (1972).
49. Bryant, M. P., Small, N., Bouma, C., Robinson, I. M.: J. Bacteriol. **76**, 529 (1958).
50. Bryant, M. P., Wolin, E. A., Wolin, R. S.: Arch. Mikrobiol. **59**, 20 (1967).
51. Bryant, M. P., McBride, B. C., Wolin, R. S.: J. Bacteriol. **95**, 118 (1968).
52. Bryant, M. P.: Amer. Chem. Soc. Abstr., Microbiol. Sect. p. 18, (1969).
53. Bryant, M. P., Campbell, L. L., Reddy, C. A., Crabill, M. R.: Appl. Environ. Microbiol. **33**, 1162 (1977).
54. Buchanan, B. B., Evans, M. C. W., Arnon, D. I.: Arch. Microbiol. **59**, 32 (1967).
55. Buchanan, R. E., Gibbons, N. E. (eds.).: Bergey's Manual of Determinative Bacteriology, 8th edition. Williams and Wilkins Co., Baltimore 1974.
56. Bulen, W. A., Burns, R. C., LeComte, J. R.: Proc. Nat' 1. Acad. Sci. **53**, 532 (1965a).
57. Bulen, W. A., LeComte, J. R., Burris, R. H., Hinkson, J.: In "Non-heme Iron Proteins" (Ed. A. San Pietro). Antioch Press, Yellow-Springs 1965b, p. 261.
58. Burke, K. A., Lascelles J.: J. Bacteriol. **123**, 308 (1965b).
59. Burns, R. C.: "Non heme Iron Proteins" (Ed. A. San Pietro), Antioch Press, Yellow-Springs, 1965, p. 289.
60. Burris, R. H., Orme-Johnson, W. H.: In "Microbial Iron Metabolism" (Ed. J. B. Neilands), Academic Press, New York, p. 187 (1976).
61. Burton, M. B., Stadtmann, E. E.: J. Biol. Chem. **202**, 873 (1953).
62. Burton, K.: Biochem. J. **59**, 44 (1955).
63. Bushnell, I. D.: J. Bacteriol. 7, 373 (1922).
64. Butlin, K. R., Adams, M. E.: Nature **160**, 154 (1947).
65. Butlin, K. R., Adams, M. E., Thomas, M.: J. Gen. Microbiol. **3**, 46 (1949).
66. Calvin, M.: Science **184**, 375 (1974).
67. Calloway, D. H.: Nature **212**, 1238 (1966).
68. Campbell, L. L., Postgate, J. R.: Bacteriol. Rev. **29**, 359 (1965).
69. Canale-Parola, E., Wolfe, R. S.: J. Bacteriol. **79**, 860 (1960).
70. Cardon, B. P., Barker, H. A.: Arch. Biochem. **12**, 165 (1947).
71. Chen, J. S., Mortenson, L. E.: Biochim. Biophys. Acta **371**, 283 (1974).
72. Chung, K.: Appl. Environ. Microbiol. **31**, 342 (1976).
73. Cille, G. C., Henzen, M. R., Stander, G. J., Baillie, R. D.: Water Res. **3**, 623 (1969).
74. Clarke, P. H.: J. Gen. Microbiol. **8**, 397 (1953).
75. Clifton, C. E.: J. Bacteriol. **39**, 485 (1940).
76. Clifton, C. E.: J. Bacteriol. **44**, 179 (1942).

77. Clowes, R. C.: J. Gen. Microbiol. **18**, 140 (1958).
78. Cole, J. A., Wimpenny, J. W. T.: Biochim. Biophys. Acta **128**, 419 (1966).
79. Cole, J. A.: Biochim. Biophys. Acta **162**, 356 (1968).
80. Cole, J. A., Wimpenny, J. W. T.: Biochim. Biophys. Acta **162**, 39 (1968).
81. Cole, J. A., Wimpenny, J. W. T.: Biochim. Biophys. Acta **162**, 39 (1969).
82. Cole, J. A., Ward, F. B.: J. Gen. Microbiol. **76**, 21 (1973).
83. Coppenet, M., Ducet, G.: Ann. Agron. **18**, 33 (1948).
84. Cowie, D. B., Bolton, E. T., Sands, M. K.: J. Bacteriol. **62**, 63 (1951).
85. Cowles, P. B., Rettger, L. F.: J. Bacteriol. **21**, 167 (1931).
86. Curtis, W., Ordal, E. J.: J. Bacteriol. **21**, 167 (1931).
87. Czerkawski, J. W., Harfoot, C. G., Breckenridge, G.: J. Appl. Bact. **35**, 537 (1972).
88. Daesch, G., Mortenson, L. E.: J. Bacteriol. **110**, 103 (1969).
89. Daesch, G., Mortenson, L. E.: J. Bacteriol. **110**, 103 (1972).
90. Dawes, E. A., Foster, S. M.: Biochim. Biophys. Acta **22**, 253 (1956).
91. Decker, K., Jungermann, K., Thauer, R. K.: Angew. Chem. Int. Ed. Engl. **9**, 138 (1970).
92. De Groot, G. N., Stouthamer, A. H.: Arch. Mikrobiol. **66**, 220 (1969).
93. De Groot, G. N., Stouthamer, A. H.: Arch. Mikrobiol. **74**, 340 (1970).
94. De Ley, J.: In "Microbial Classification", 12th Symp. Soc. Gen. Microbiol., Cambridge Univ. Press, Cambridge 1962, p. 164.
95. De Vries, W., van Wijck-Kapteyn, W. M. C., Oosterhuis, S. K. H.: J. Gen. Microbiol. **81**, 69 (1974).
96. Deyhle, R. R., Barton, L. L.: Can. J. Microbiol. **23**. 125 (1977).
97. Dixon, R. O. D.: Ann. Bot. **31**, 179 (1967).
98. Dixon, R. O. D.: Arch. Mikrobiol. **62**, 272 (1968).
99. Dixon, R. O. D.: Rhizobia. Ann. Rev. Microbiol. **23**, 137 (1969).
100. Douglas, F., Rigby, G. J.: J. Appl. Bacteriol. **37**, 251 (1974).
101. Douglas, M. W., Ward, F. B., Cole, J. A.: J. Gen. Microbiol. **80**, 557 (1974).
102. Egami, F., Ishimoto, M., Taniguichi, S.: In "Haematin Enzymes" (Ed. J. E. Falk, R. Lemberg, R. K. Morton), Pergamon, Oxford 1961, p. 392.
103. Eksertsev, V. A.: Geochem. 432 (1960).
104. Enebo, L., Pherson, S. O.: Acta Polytech. Scand. **AP281**, 1 (1960).
105. Enoch, H. G., Lester, R. L.: J. Bacteriol. **110**, 1032 (1972).
106. Enoch, H. G., Lester, R.: Biochem. Biophys. Res. Commun. **61**, 1234 (1974).
107. Enoch, H. G., Lester, R. L.: J. Biol. Chem. **200**, 6693 (1975).
108. Fischer, F., Lieske, R., Winzer, K.: Biochems. Z **236**, 247 (1931).
109. Foubert, E. L., Douglas, H. C.: J. Bacteriol. **56**, 25 (1948).
110. Frankland, P. F., Frew, W.: J. Chem. Soc. **59**, 81 (1891).
111. Frankland, P. F., Lumsden, J. S.: J. Chem. Soc. **59**, 432 (1892).
112. Frankland, P. F., Macgregor, J.: J. Chem. Soc. **59**, 737 (1892).
113. Fredericks, W. W., Stadtman, E. R.: J. Biol. Chem. **240**, 4065 (1965).
114. Frenkel, A. W.: Arch. Bioch. Biophys. **38**, 219 (1952).
115. Fromageot, C., Senez, J. C.: In "Comparative Biochemistry" vol. 1, (Ed. M. Florkin, H. S. Mason), Academic Press, New York 1960, p. 347.
116. Fuchs, A. R., Bonde, G. J.: J. Gen. Microbiol. **16**, 330 (1957).
117. Fukuyama, T., Ordal, E. J.: J. Bacteriol. **90**, 673 (1965).
118. Gaffron, H., Rubin, J.: J. Gen. Physiol. **26**, 219 (1942).
119. Gest, H., Kamen, M. D.: J. Bacteriol. **58**, 239 (1949b).
120. Gest, H., Kamen, M. D.: Science **109**, 558 (1949a).
121. Gest, H.: Bacteriol. Revs. **15**, 183 (1951).
122. Gest, H.: J. Bacteriol. **63**, 111 (1952).
123. Gest, H.: Bacteriol. Revs. **18**, 43 (1954).
124. Gest, H., Peck, H. D.: J. Bacteriol. **70**, 326 (1955).
125. Gest, H., Ormerod, J. G., Ormerod, K. S.: Arch. Biochem. Biophys. **97**, 21 (1962).
126. Gest, H.: Adv. Microb. Physiol. **7**, 243 (1972).

127. Gogotov, I. N., Kosijak, A. B.: Mikrobiologia **45**, 586 (1976).
128. Golueke, G., McGauhey, P. M.: Ann. Rev. Energy **1**, 257 (1976).
129. Gottschalk, G.: Euro. J. Biochem. **5**, 346 (1968).
130. Gottwald, M., Andreesen, J. R., LeGall, J., Ljungdhal, L. G.: J. Bacteriol. **122**, 325 (1975).
131. Gray, F. H., Wilson, P. W.: J. Bacteriol. **83**, 490 (1962).
132. Gray C. T., Wimpenny, J. W. T., Hughes, D. E., Ranlett, M.: Biochim. Biophys. Acta **67**, 156 (1963).
133. Gray, C. T., Gest, H.: Science **148**, 186 (1965).
134. Gregory, D. P., Ng, D. Y., Long, G. M.: In "Electrochemistry of Cleaner Environments" (Ed. J. O. Bockris) Plenum Press, New York 1972, p. 226.
135. Green, D. E., Stickland, L. H.: Biochem. J. **28**, 898 (1934).
136. Green, M., Wilson, P. W.: J. Bacteriol. **65**, 511 (1953).
137. Grey, E. C.: Proc. Roy, Soc. **91**, 294 (1920).
138. Guarraia, L. J. Laishley, E. J., Forget, N., Peck, H. D.: Bacteriol. Proc. **P129**, 133 (1968).
139. Haddock, B. A., Kendall-Tobias, M. W.: Biochim. J. **152**, 655 (1975).
140. Haddock, B. A., Jones, C. W.: Bacteriol. Rev. **41** (1977).
141. Hamilton, R. D., Wolfe, R. S.: J. Bacteriol. **78**, 253 (1959).
142. Hansen, R. G., Henning, U.: Biochim. Biophys. Acta **122**, 355 (1966).
143. Harden, A.: J. Chem. Soc. **79**, 610 (1901).
144. Harden, A., Walpole, G. S.: Proc. Roy. Soc. **77B**, 399 (1906).
145. Harden, A., Norris, D.: Proc. Roy. Soc. **84**, 492 (1912).
146. Hardy, R. W. F., Knight, E.: Biochim. Biophys. Acta **122**, 520 (1966).
147. Hardy, R. W. F., Holsten, R. D., Jackson, E. K., Burns, R. C.: Plant Physiol. **43**, 1185 (1968).
148. Hardy, R. W. F., Knight, E.: In "Phytochemistry" Vol. 1 (Ed. L. Reinhold, Y. Linschitz). Interscience Publishers, New York 1968, p. 407.
149. Hardy, R. W. F., Burns, R. C., Holsten, R. D.: Soil Biol. Biochem. **5**, 47 (1973).
150. Harris, J. E.: J. Path. and Bact. **23**, 30 (1919).
151. Hatfield, W. D.: Ind. Eng. Chem. **20**, 174 (1928).
152. Healey, F. P.: Planta **91**, 220 (1970a).
153. Healey, F. P.: Plant Physiology **45**, 153 (1970b).
154. Henning, U.: Biochem. Z **337**, 490 (1963).
155. Hernandez, E., Johnson, M. J.: J. Bacteriol. **94**, 991 (1967).
156. Hillmer, P., Gottschalk, G.: FEBS Lett. **21**, 351 (1972).
157. Hillmer, P., Gest, H.: J. Bacteriol. **129**, 724 (1977a).
158. Hillmer, P., Gest, H.: J. Bacteriol. **129**, 732 (1977b).
159. Hino, S.: J. Biochem. (Tokyo) **47**, 482 (1960).
160. Hoare, D. S., Hoare, S. L.: J. Bacteriol. **100**, 1124 (1969).
161. Hoberman, H. D., Rittenberg, D.: J. Biol. Chem. **147**, 211 (1943).
162. Hoch, G. E., Schneider, K. C., Burris, R. H.: Biochim. Biophys. Acta **37**, 273 (1960).
163. Hoogerheide, J. C., Kocholaty, W.: Biochem. J. **32**, 949 (1938).
164. Hoover, S. R.: Science **183**, 824 (1974).
165. Hoppe-Seyler, F.: Pflugers Arch. Ges. Physiol. **12**, 1 (1876).
166. Humphrey, A. E.: Chem. Eng. **81**, 98 (1974).
167. Hungate, R. E.: Bacteriol. Rev. **14**, 1 (1950).
168. Hungate, R. E.: Can. J. Microbiol. **3**, 289 (1957).
169. Hungate, R. E.: In "The Rumen and Its Microbes" (Ed. R. E. Hungate), Academic Press Inc., New York 1966, p. 273.
170. Hungate, R. E.: Arch. für Mikrobiol. **59**, 158 (1967).
171. Hungate, R. E.: ASM News **40**, 833 (1974).
172. Hyndman, L. A., Burris, R. H., Wilson, P. W.: J. Bacteriol. **65**, 522 (1953).
173. Iannotti, E. L., Kafkewitz, D., Wolin, M. J., Bryant, M. P.: J. Bacteriol. **114**, 1231 (1973).
174. Inderlied, C. G., Delwiche, E. A.: J. Bacteriol. **114**, 1206 (1973).
175. Ishimoto, M., Yagi, T., Shiraki, M.: J. Biochem. (Tokyo) **44**, 707 (1957).
176. Ishimoto, M., Yagi, T.: J. Biochem. (Tokyo) **49**, 103 (1961).

177. Johns, A. T.: J. Gen. Microbiol. **5**, 317 (1951a).
178. Johns, A. T.: J. Gen. Microbiol. **5**, 326 (1951b).
179. Johns, A. T., Barker, H. A.: J. Bacteriol. **80**, 837 (1960).
180. Johnson, M. J., Peterson, W. H., Fred, E. B.: J. Biol. Chem. **91**, 569 (1931).
181. Johnson, P. A., Quayle, J. R.: Biochem. J. **93**, 281 (1964).
182. Joyner, A. E., Winter, W. T., Godbout, D. M.: Can. J. Microbiol. **23**, 346 (1977).
183. Jungermann, K., Thauer, R. K., Rupprecht, E., Ohrloff, C., Decker, K.: FEBS Lett. **3**, 144 (1969).
184. Jungermann, K., Kirchniawy, H., Thauer, R. K.: Biochem. Biophys. Res. Commun. **41**, 682 (1970).
185. Jungermann, K., Rupprecht, E., Ohrloff, C., Thauer, R. K., Decker, K.: J. Biol. Chem. **246**, 960 (1971).
186. Jungermann, K., Thauer, R. K., Leimenstoll, G., Decker, K.: Biochim. Biophys. Acta **305**, 268 (1973).
187. Kalmitsky, G., Werkmann, C. H.: Arch. Biochem. **2**, 113 (1943).
188. Kaltwasser, H., Stuart, T. S., Gaffron, H.: Planta **89**, 309 (1969).
189. Karczag, L., Schiff, E.: Biochem. Zeitschr. **70**, 325 (1915).
190. Kearney, J. J., Sagers, R. D.: J. Bacteriol. **109**, 152 (1972).
191. Keister, D. L.: J. Bacteriol. **123**, 1265 (1975).
192. Kemp, J. D., Atkinson, D. E.: J. Bacteriol. **92**, 628 (1966).
193. Kempner, W., Kubowitz, F.: Biochem. Zeitschr. **265**, 245 (1933).
194. Kennedy, I. R., Morris, J. A., Mortenson, L. E.: Biochim. Biophys. Acta **153**, 777 (1968).
195. Kessler, E.: Dtsch. Bot. Ges. (N. F.) **1**, 92 (1962).
196. Kessler, E.: Arch. Mikrobiol. **55**, 346 (1967).
197. Kessler, E., Czygan, F. C., Fott, B., Novakova, M.: Arch. Protistenk **110**, 462 (1968).
198. Kessler, E., Zweier, I.: Arch. Mikrobiol. **79**, 44 (1971).
199. Kessler, E.: Arch. Mikrobiol. **93**, 91 (1973).
200. Khosrovi, B. R., MacPherson, R., Miller, J. A. R.: Arch. Mikrobiol. **80**, 324 (1971).
201. Khouvine, Y.: Ann. Inst. Pasteur, **37**, 711 (1923).
202. Kinsky, S. C., Stadtman, E. R., McLoyd, H. K.: J. Biol. Chem. **236**, 574 (1961).
203. Klein, S. M., Sagers, R. D.: J. Biol. Chem. **241**, 197 (1966).
204. Kluyver, A. J.: Chemical Activities of Microorganisms. Univ. Press, London 1931.
205. Kluyver, A. J.: In "Intern. Congr. Microbiol." 6th Congr. Rome (1953), p. 71.
206. Knappe, J., Schacht, J., Mockely, M., Hopner, T., Vetter, H., Edenharder, R.: Eur. J. Biochem. **11**, 316 (1969).
207. Knappe, J., Blaschkowski, H. P., Grobner, P., Schmitt, T.: Euro. J. Biochem. **50**, 253 (1974).
208. Knight, E., Hardy, R. W. F.: J. Biol. Chem. **241**, 2752 (1966).
209. Knoell, H., Knappe, J.: Euro. J. Biochem. **50**, 245 (1974).
210. Knook, D. L., Riet, J., Van, T., Planta, R. J.: Biochim. Biophys. Acta **292**, 236 (1973).
211. Koch, B., Evans, H. G., Russell, S.: Proc. Nat' 1. Acad. Sci. **58**, 1343 (1967).
212. Koepsell, H. J., Johnson, M. J.: J. Biol. Chem. **145**, 379 (1942).
213. Koepsell, H. J., Johnson, M. J., Lipmann, F.: See Lipmann, F. Ann. Rev. Biochem. **12**, 1 (1943).
214. Koepsell, H. J., Johnson, M. J., Meek, J. S.: J. Biol. Chem. **154**, 535 (1944).
215. Kohlmiller, E. F., Gest, H.: J. Bacteriol. **61**, 269 (1951).
216. Kondratieva, E. N., Gogotov, J. N.: Nature **221**, 83 (1969).
217. Kondratieva, E. N., Malofeeva, I. V., Sumarukova, R. S.: Mikrobiologiya **38**, 13 (1969).
218. Kondratieva, E. N.: In "Microbial Energy Conversion" (Ed. H. G. Schlegel and J. Barnea), Seminar held in Göttingen, 4–8th Oct., Göttingen 1976, p. 205.
219. Kondratieva, E. N., Gogotov, Z. N.: Izvestija Akad. Nauk SSSR, Ser. Biol. **69** (1976).
220. Kornberg, H. L., Elsden, S. R.: Adv. Enzymol. **23**, 401 (1961).
221. Kornberg, H. L., Collins, J. F., Bigley, D.: Biochim. Biophys. Acta **39**, 9 (1960).
222. Koyama, T.: J. Geophys. Res. **68**, 3971 (1963).
223. Kroger, A., Schimkat, M., Niedermaier, S.: Biochim. Biophys. Acta **347**, 273 (1974).

224. Kubowitz, F.: Biochem. Zeitschr. **274**, 285 (1934).
225. Kurata, Y.: Exp. Cell Res. **28**, 424 (1962).
226. Kushner, D. J., Quastel, J. H.: Proc. Soc. Exptl. Biol. Med. **82**, 388 (1953).
227. Lawrence, A. W.: In "Advances in Chemistry" (Ed. R. F. Gould) series 105, American Chemical Society, Washington 1971, p. 163.
228. Lee, S. B., Wilson, P. W.: J. Biol. Chem. **151**, 377 (1943).
229. LeGall, J., Dragoni, N.: Biochem. Biophys. Res. Commun. **23**, 145 (1966).
230. LeGall, J., Postgate, J. R.: Adv. Microb. Res. **10**, 82 (1973).
231. Lester, R. L., DeMoss, J. A.: J. Bacteriol. **105**, 1006 (1971).
232. Lewis, A. J., Miller, J. D. A.: J. Gen. Microbiol. **90**, 286 (1975).
233. Li, L. F., Ljungdahl, L., Wood, H. G.: J. Bacteriol. **92**, 405 (1966).
234. Lichstein, H. C., Boyd, R. B.: J. Bacteriol. **62**, 415 (1951).
235. Lichstein, H. C., Boyd, R. B.: Proc. Soc. Exptl. Biol. Med. **79**, 308 (1952a).
236. Lichstein, H. C., Boyd, R. B.: Bacteriol. Proc. 1952, 161 (1952b).
237. Lichstein, H. C., Boyd, R. B.: J. Bacteriol. **65**, 617 (1953).
238. Lindmark, D. G., Muller, M.: J. Biol. Chem. **248**, 7724 (1973).
239. Lipmann, F.: Cold Spring Harbor Symp. on Quant. Biol. **7**, 248 (1939).
240. Lipmann, F.: Adv. Enzymol. **1**, 99 (1941).
241. Lipmann, F., Tuttle, L. S.: J. Biol. Chem. **158**, 505 (1945).
242. Ljones, T., Burrins, R. H.: Biochim. Biophys. Acta **275**, 93 (1972).
243. Ljungdahl, L., Wood, H. G.: Ann. Rev. Microbiol. **23**, 515 (1969).
244. Losada, M., Nozaki, M., Arnon, D. I.: In "Light and Life", (Ed. W. D. McElroy, B. Glass), John Hopkins Press, Baltimore 1961, p. 570.
245. Lovenberg, W.: In "Microbial Iron Metabolism" (Ed. J. B. Neilands), Academic Press, New York, p. 161.
246. MacGregor, C. H.: J. Bacteriol. **121**, 1111 (1975).
247. Macy, J., Kulla, H., Gottschalk, G.: J. Bacteriol. **125**, 423 (1976).
248. Mahler, H. R., Cordes, E. H.: In „Biological Chemistry", Harper and Row, New York 1966, p. 437.
249. McBride, B. C., Wolfe, R. S.: In „Advances in Chemistry" (Ed. R. F. Gould) series 105, American Chemical Society, Washington 1971, p. 11.
250. McCarty, P. L.: In "Advances in Chemistry" (Ed. R. F. Gould), series 105, American Chemical Society, Washington 1971, p. 91.
251. Mechalas, B. J., Rittenberg, S. C.: J. Bacteriol. **80**, 501 (1960).
252. Michel, J. W.: 166th A. C. S. Nat' 1. Meet., Div. Fuel Chem. **18**, 1 (1973).
253. Mickelson, M. N., Werkman, C. H.: J. Bacteriol. **37**, 619 (1939).
254. Miller, P. L.: Contrib. Boyce Thompson Inst. **16**, 78 (1950).
255. Miller, T. L., Wolin, M. L.: J. Bacteriol. **116**, 836 (1973).
256. Millet, J.: Compt. Rend. **240**, 253 (1954).
257. Mitchell, P.: Biochem. Soc. Transact. **4**, 399 (1976).
258. Mortenson, L. E., Wilson, P. W.: J. Bacteriol. **62**, 513 (1951).
259. Mortenson, L. E., Valentine, R. C., Carnaham, J. E.: Biochem. Biophys. Res. Commun. **7**, 448 (1962).
260. Mortenson, L. E.: Ann. Rev. Microbiol. **17**, 115 (1963).
261. Mortenson, L. E.: Surv. Progr. Chem. **4**, 127 (1968).
262. Mortenson, L. E.: Biochem. Soc. Transact. **1**, 35 (1973).
263. Mortenson, L. E., Nakos, G.: In "Molecular Biology, An International Series of Monographs and Textbooks", (Ed. W. Lovenberg), "Iron and Sulfur Proteins", Vol. I, Academic Press, New York 1973.
264. Mortenson, L. E., Chen, J.: In "Microbial Iron Metabolism" (Ed. J. B. Neilands), Academic Press, New York 1974, p. 231.
265. Mortenson, L. E., Nakos, J. S.: In "Microbial Iron Metabolism" (Ed. J. B. Neilands), Academic Press, New York 1974.
266. Mortlock, R. P., Valentine, R. C., Wolfe, R. S.: J. Biol. Chem. **234**, 1653 (1959).

267. Muller, M.: Ann. Rev. Microbiol. **29**, 467 (1975).
268. Nagy, A., Kari, C., Hernadi, F.: Arch. Mikrobiol. **65**, 391 (1969).
269. Nakamura, H.: Acta. Phytochim. (Japan) **11**, 109 (1939).
270. Nakamura, H.: Acta. Phytochim. (Japan) **12**, 43 (1941).
271. Nakamura, T., Sato, R.: Nature **185**, 163 (1960).
272. Nakayama, H., Midwinter, G. G., Krampitz, L. O.: Arch. Biochem. Biophysics **143**, 526 (1971).
273. Nakos, G., Mortenson, L. E.: Biochim. Biophys. Acta **227**, 576 (1971a).
274. Nakos, G., Mortenson, L. E.: Biochemistry **10**, 2442 (1971b).
275. Nason, A.: Bacteriol. Rev. **26**, 16 (1962).
276. Neilson, A. H., Nordlund, S.: J. Gen. Microbiol. **91**, 53 (1975).
277. Neiman, O.: Bacteriol. Rev. **18**, 147 (1954).
278. Neish, A. C., Blackwood, A. C., Robertson, F. M., Ledingham, G. A.: Can. J. Res. **26**, 335 (1948).
279. Newton, J. W.: Science **191**, 559 (1976).
280. Nojiri, T., Tanaka, F., Nakayama, I.: J. Biochem. **69**, 789 (1971).
281. O'Brien, R. W., Morris, J. G.: Arch. Mikrobiol. **84**, 225 (1972).
281a. Ohwaki, K., Hungate, R. E.: Appl. Environ. Microbiol. **33**, 1270 (1977).
282. Omelianski, W.: Compt. Rend. Acad. Sci. **125**, 970 (1897).
283. Omelianski, W.: Zentr. Bakt. Abt. II, Vol. 8 (1902).
284. Ordal, E. J., Halvorson, H. O.: J. Bacteriol. **38**, 199 (1939).
285. Orme-Johnson, W. H.: Ann. Rev. Biochem. **42**, 159 (1973).
286. Ormerod, J. G., Ormerod, K. S., Gest, H.: Arch. Biochem. Biophys. **94**, 449 (1961).
287. Ormerod, J. G., Gest, H.: Bacteriol. Rev. **26**, 51 (1962).
288. Osburn, O. L., Brown, R. W., Werkman, C. H.: Iowa State Coll. J. Sci. **12**, 275 (1938).
289. Oshchepkov, V. P., Krasnovskii, A. A.: Sov. Plant Physiol. **19**: 931 (1972). Engl. translation Fiziol Rast. **19**, 1090 (1973).
290. Oshchepkov, V. P., Nikitina, K. A., Gusev, M. V., Krasnovskii, A. A.: Dokl. Akad. Nauk., USSR, Ser. Biol. **213**, 739 (1973).
291. Oshchepkov, V. P., Krasnovskii, A. A.: Isvestija Adak. Nauk USSR, Serija Biol. **87** (1976).
292. Oster, M. O., Wood, N. P.: J. Bacteriol. **87**, 104 (1964).
293. Packer, L., Vishniac, W.: Biochem. Biophys. Acta **17**, 153 (1955).
294. Packer, L.: Arch. Biochem. Biophys. **78**, 54 (1958).
295. Pakes, W. C. C., Jollyman, W. H.: J. Chem. Soc. **79**, 386 (1901).
296. Pappenheimer, A. M., Shaskan, E.: J. Biol. Chem. **155**, 265 (1944).
297. Peck, H. D., Gest, H.: Biochim. Biophys. Acta **15**, 587 (1944).
298. Peck, H. D., Gest, H.: J. Bacteriol. **71**, 70 (1956).
299. Peck, H. D., Gest, H.: J. Bacteriol. **73**, 706 (1957a).
300. Peck, H. D., Gest, H.: J. Bacteriol. **73**, 569 (1957b).
301. Peck, H. D.: Proc. Nat. Acad. Sci. **45**, 701 (1959).
302. Peck, H. D.: Bacteriol. Proc. **60**, 167 (1960).
303. Peck, H. D.: J. Biol. Chem. **237**, 198 (1962).
304. Peck, H. D.: Biochem. Biophys. Res. Commun. **22**, 112 (1966).
305. Peck, H. D.: Ann. Rev. Microbiol. **22**, 489 (1968).
306. Pederson, C. S., Breed, R. S.: J. Bacteriol. **16**, 163 (1928).
307. Peel, J. L., Barker, H. A.: Biochem. J. **53**, xxix (1953).
308. Pfennig, F.: Ann. Rev. Microbiol. **21**, 311 (1967).
309. Pfennig, S., Biebl, H.: Arch. Microbiol. **110**, 3 (1976).
310. Pichinoty, F.: Biochim. Biophys. Acta **64**, 111 (1962).
311. Pine, L., Haas, V., Barker, H. A.: J. Bacteriol. **68**, 227 (1954).
312. Pine, M. J.: In "Advances in Chemistry" (Ed. R. F. Gould) series 105, American Chemical Society, Washington 1971, p. 1.
313. Pinsent, J.: Biochem. J. **57**, 10 (1954).
314. Pinsky, M. J., Stokes, J. L.: J. Bacteriol. **64**, 151 (1952).

315. Popoff, L.: Arch. F. D. Ges. Physiol. **10**, 113 (1875).
316. Post, R. F., Ribe, F. L.: Science **186**, 397 (1974).
317. Postgate, J. R.: J. Gen. Microbiol. **5**. 725 (1951).
318. Postgate, J. F.: Research (London) **5**, 198 (1952).
319. Postgate, J. R.: J. Gen. Microbiol. **15**, 186 (1956).
320. Postgate, J. R.: Z. Allgem. Mikrobiol. **1**, 53 (1960).
321. Postgate, J. R.: In "The Haematin Enzymes" (Ed. J. E. Falk, R. Lemberg, R. K. Morton), Part 2, Pergamon Press, Oxford 1961, p. 407.
322. Postgate, J. R.: J. Gen. Microbiol. **30**, 481 (1963).
323. Postgate, J. R.: Bacteriol. Rev. **28**, 425 (1965).
324. Postgate, J. R.: Lab. Pract. **15**, 1239 (1966).
325. Postgate, J. R., Cambell, L. L.: Bacteriol. Rev. **30**, 732 (1966).
326. Prazmowski, A.: Untersuchungen über die Entwicklungsgeschichte und Fementwirkung einiger Bacterien-Arten, Leipzig 1880, p. 58.
327. Pretorius, W. A.: Water Res. **5**, 681 (1971).
328. Quadri, S. M. H., Hoare, D. S.: J. Bacteriol. **95**, 2444 (1968).
329. Quastel, J. H., Whetham, M. D.: Biochem. J. **19**, 520, 645–651 (1925).
330. Radmer, R., Kok, B.: Ann. Rev. Biochem. **44**, 409 (1975).
331. Raeburn, S., Rabinowitz, J. C.: Biochem. Biophys. Res. Commun. **18**, 303 (1965).
332. Rao, K. K., Rosa, L., Hall, D. O.: Biochem. Biophys. Res. Commun. **68**, 21 (1976).
333. Reddy, C. A., Bryant, M. P., Wolin, M. J.: J. Bacteriol. **110**, 126 (1972a).
334. Reddy, C. A., Bryant, M. P., Wolin, M. J.: J. Bacteriol. **110**. 133 (1972b).
335. Reed, R. M.: Trans Amer. Inst. Chem. Eng. **41**, 453 (1945).
336. Reed, L. J.: In "Comparative Biochemistry" (Ed. M. Florkin, E. H. Stotz), Vol. 14, Elsevier Publ. Co., Amsterdam 1966, p. 99.
337. Reed, L. J., Cox, D. J.: In "The Enzymes" (Ed. P. D. Boyer), Vol. 1, Academic Press Inc., New York 1970, p. 213.
338. Reeves, S. G., Rao, K. K., Rosa, L., Hall, D. O.: In "Microbial Energy Conversion" (Ed. H. C. Schlegel, J. Barnea), Seminar held in Göttingen 4–8th October, 1976. E. Goltze, Göttingen 1976, p. 235.
339. Renwick, G. M., Giumarro, C., Siegel, S. M.: Plant Physiol. **39**, 303 (1964).
340. Reynolds, H., Werkman, C. H.: J. Bacteriol. **33**, 603 (1937).
341. Roberts, R. B., Alberson, P. H., Cowie, D. B., Bolton, E. T. Britten, R. J.: In "Studies of biosynthesis in Escherichia coli" (Ed. R. B. Roberts *et al.*), Carnegie Institute of Wash. Inc., Washington 1964.
342. Robinson, J. R., Sagers, R. D.: J. Bacteriol. **112**, 465 (1972).
343. Robson, G. R.: Science **184**, 371 (1974).
344. Rogosa, M.: J. Bacteriol. **87**, 162 (1964a).
345. Rogosa, M.: J. Bacteriol. **87**, 574 (1964b).
346. Rogosa, M., Bishop, F. S.: J. Bacteriol. **88**, 37 (1964).
347. Rose, D. J.: Science **184**, 351 (1974).
348. Rose, I. A., Grunberg-Manago, M., Korey, S. R., Ochoa, S.: J. Biol. Chem. **211**, 737 (1954).
349. Ruiz-Herrera, J., Alvarex, A., Ant. van Leeuwen.: J. Microbiol. Serol. **38**, 479 (1972).
350. Ruiz-Herrera, J., DeMoss, J. A.: J. Bacteriol. **99**, 720 (1969).
351. Ruiz-Herrera, J., Alvarez, A., Figueroa, I.: Biochim Biophys. Acta **289**, 254 (1972).
352. San Pietro, A.: In "Microbial Energy Conversion" (Ed. H. C. Schlegel, J. Barnea), Seminar held in Göttingen 4–8th October, 1976. E. Goltze, Göttingen 1976, p. 217.
353. Scheifinger, C. C., Linehan, B., Wolin, M. J.: Appl. Microbiol. **29**, 480 (1975).
354. Schick, H. J.: Arch. Mikrobiol. **75**, 102 (1971).
355. Schlossman, K., Lynen, F.: Biochem. Z. **238**, 591 (1957).
356. Schoberth, S., Gotschalk, G.: Arch. Mikrobiol. **65**, 318 (1969).
357. Schon, G.: Arch. Mikrobiol. **79**, 147 (1971).
358. Schon, G., Bidermann, M.: Arch. Mikrobiol. **85**, 77 (1972).
359. Schon, G., Bidermann, M.: Biochim. Biophys. Acta **304**, 65 (1973).

359a. Schon, G., Voelskow, H.: Arch. Microbiol. **107**, 87 (1976).
360. Schreckenbach, T.: In "Microbial Energy Conversion" (Ed. H. G. Schlegel, J. Barnea), Seminar held in Göttingen, 4–8th October, 1976. E. Goltze, Göttingen 1976, p. 245
361. Senn, J. B.: Ph. D. Thesis, Univ. of Calif., Los Angeles 1946.
362. Seto, B., Mortenson, L. E.: Biochem. Biophys. Res. Commun. **53**, 419 (1973).
363. Seto, B., Mortenson, L. E.: J. Bacteriol. **117**, 805 (1974).
364. Shaumugam, K. T., Buchanan, B. B., Arnon, D. I.: Biophys. Acta **256**, 477 (1972).
365. Shethna, Y. I., Stombaugh, N. A., Burris, R. H.: Biochem. Biophys. Res. Commun. **42**, 1108 (1971).
366. Shimizu, M., Suzuki, T., Kameda, K., Abiko, Y.: Biochim. Biophys. Acta **191**, 550 (1969).
367. Shug, A. L., Wilson, P. W., Green, D. E., Mahler, H. R.: J. Amer. Chem. Soc. **76**, 3355 (1954).
368. Shug, A. L., Wilson, P. W.: Federation Proc. **15**, 335 (1956).
369. Shum, A. C., Murphy, J. C.: J. Bacteriol. **110**, 447 (1972).
370. Sjolander, N. O.: J. Bacteriol. **34**, 419 (1937).
371. Skulachev, V. P.: FEBS Lett. **64**, 23 (1976).
372. Smit, J.: J. Path. Bacteriol. **36**, 455 (1933).
373. Sokatch, L. R.: Bacterial Physiology and Metabolism, Academic Press, London 1969.
374. Sorokin, Y. J.: Nature (London) **210**, 551 (1966).
375. Speakman, H. B.: J. Biol. Chem. **43**, 401 (1920).
376. Speck, M. L., Stark, C. N.: J. Bacteriol. **44**, 687 (1942).
377. Spruit, C. J. P.: Plant Physiol. **35**, 988 (1958).
378. Stadtman, E. R., Barker, H. A.: J. Biol. Chem. **180**, 1169 (1949a).
379. Stadtman, E. R., Barker, H. A.: J. Biol. Chem. **180**, 1168 (1949b).
380. Stadtman, E. R., Barker, H. A.: J. Biol. Chem. **184**, 769 (1950).
381. Stadtman, T. C., Barker, H. A.: J. Bacteriol. **62**, 269 (1951).
382. Stadtman, E. R.: In "Methods in Enzymology" (Ed. S. P. Colowick, N. O. Kaplan) vol. 1, Academic Press Inc., New York 1955, p. 596.
383. Stadtman, E. R., Burton, R. M.: In "Methods in Enzymology" (Ed. S. P. Colowick and N. O. Kaplan) vol. 1, Academic Press Inc., New York 1955, p. 518.
384. Stadtman, E. R.: In "Biochemical Energetics" (Ed. N. O. Kaplan, E. P. Kennedy), Academic Press, New York 1966, p. 36.
385. Stadtman, T. C.: Ann. Rev. Microbiol. **21**, 121 (1967).
386. Stahly, G. L.: Iowa State Coll. J. Sci **11**, 110 (1936).
387. Stanier, R. Y.: Bacteriol. Rev. **25**, 1 (1961).
388. Stanier, R. Y., Adelberg, E. A., Ingraham, J.: The Microbial World, 4th Edition, Prentice-Hall, New Jersey 1976.
389. Starkey, R. L.: Ant. van Leuwen, J. Micro. Serol. **12**, 193 (1947).
390. Stephenson, M., Stickland, L. H.: Biochem. J **25**, 205 (1931a).
391. Stephenson, M., Stickland, L. H.: Biochem. J. **25**, 215 (1931b).
392. Stephenson, M., Stickland, L. H.: Biochem. J. **26**, 712 (1932).
393. Stephenson, M., Stickland, L. H.: Biochem. J. **27**, 1517 (1933a).
394. Stephenson, M., Stickland, L. H.: Biochem. J. **27**, 1528 (1933b).
395. Stephenson, M.: Ergebn. Enzymforsch. **6**, 139 (1937).
396. Stewart, W. D. P., Pearson, H. W.: Proc. Roy. Soc. **B175**, 293 (1970).
397. Stickland, L. H.: Biochem. J. **23**, 1187 (1929).
398. Stokes, J. L.: J. Bacteriol. **57**, 147 (1949).
399. Stormer, K.: Zentr. Bakt. Abt. II, **13**, 35, 171, 306 (1904).
400. Stouthamer, A. H., Bettenhausen, C., Van Hartingsveldt, J., Van't Riet, J., Planta, R. J.: Arch. Mikrobiol. **58**, 228 (1967).
401. Stouthamer, A. H.: Adv. Microb. Physiol. **14**, 315 (1976).
402. Strandberg, G. W., Wilson, P.: Can. J. Microbiol. **14**, 25 (1968).
403. Strecker, H. J.: J. Biol. Chem. **189**, 815 (1951).
404. Streicher, S. L., Valentine, R. C.: Ann. Rev. Biochem. **42**, 279 (1973).
405. Stuart, T. S., Kaltwasser, H.: Planta **91**, 302 (1970).

406. Stuart, J. S.: Planta **96**, 81 (1971).
407. Sumper, M., Herrmann, G.: FEBS Lett. in press 1976.
408. Sykes, R. M.: Ph. D. Thesis, Purdue University, Lafayette 1970.
409. Sykes, R. M., Kirch, E. J.: Dev. Ind. Microbiol. **11**, 357 (1970).
410. Taber, W. A.: Ann. Rev. Microbiol. **30**, 263 (1976).
411. Tagawa, K., Arnon, D. I.: Nature (London) **195**, 537 (1962).
412. Tarr, H. L. A.: Biochem. J. **27**, 759 (1933).
413. Tasman, A., Pot, A. W.: Biochem. J. **29**, 1749 (1935).
414. Tasman, A.: Biochem. J. **29**, 2446 (1935).
415. Taylor, G. T.: Process. Biochem. **29** (1975).
416. Thauer, R. K., Jungermann, K., Henninger, H., Wenning, J., Decker, K.: Euro. J. Biochem. **4**, 173 (1968).
417. Thauer, R. K., Jungermann, K., Rupprecht, E., Decker, K.: FEBS Lett. **4**, 108 (1969).
418. Thauer, R. K., Rupprecht, E., Jungermann, K.: FEBS Lett. **8**, 304 (1970a).
419. Thauer, R. K. Rupprecht, E., Jungermann, K.: FEBS Lett. **9**, 271 (1970b).
420. Thauer, R. K.: FEBS Lett. **27**, 111 (1972).
421. Thauer, R. K., Kirchniawy, F. H., Jungermann, K. S.: Euro. J. Biochem. **27**, 282 (1972).
422. Thauer, R. K., Fuchs, G., Jungermann, K. S.: J. Bacteriol. **118**, 754 (1974).
423. Thauer, R. K., Kaufer, B., Fuchs, G.: Eur. J. Biochem. **55**, 111 (1975).
424. Thauer, R.: In "Microbial Energy Conversion" (Ed. H. G. Schleger, J. Barnea), Seminar held in Göttingen, 4–8th October, 1976. E. Goltze, Göttingen 1976, p. 201.
425. Thauer, R., Jungermann, K., Decker, K.: Bacteriol. Rev. **41**, 100 (1977).
426. Thiel, P. G.: Water Res. **3**, 215 (1969).
427. Thorne, K. J. I., Jones, M. E.: J. Biol. Chem. **238**, 2992 (1963).
428. Todd, L. C.: J. Infect. Dis. **20**, 151 (1917).
429. Tsuchiya, T.: J. Bacteriol. **129**, 763 (1977).
430. Twarog, R., Wolfe, R. S.: J. Bacteriol. **86**, 112 (1963).
431. Tzeng, S. F., Wolfe, R. S., Bryant, M. P.: J. Bacteriol. **121**, 184 (1975a).
432. Tzeng, S. F., Bryant, M. P., Wolfe, R. S.: J. Bacteriol. **121**, 192 (1975b).
433. Uffen, R. L., Sybesma, C., Wolfe, R. S.: J. Bacteriol. **108**, 1348 (1971).
434. Umbreit, W. W.: In "The Enzymes" (Ed. J. B. Summer, K. Myrback), Vol. II, Part I, 1951, p. 329.
435. Upadhyay, J., Stokes, J. L.: J. Bacteriol. **85**, 177 (1963).
436. Utter, M. F., Werkman, C. H.: Arch. Biochem. **2**, 491 (1943).
437. Utter, M. F., Werkman, C. H.: Arch. Biochem. **5**, 413 (1944).
438. Uyeda, K., Rabinowitz, J. C.: J. Biol. Chem. **246**, 3120 (1971a).
439. Uyeda, K., Rabinowitz, J. C.: J. Biol. Chem. **246**, 3111 (1971b).
440. Valentine, R. C., Jackson, R. L., Wolfe, R. S.: Biochem. Biophys. Res. Commun. **7**, 453 (1962).
441. Valentine, R. C., Wolfe, R. S.: J. Bacteriol. **85**, 1114 (1963).
442. Valentine, R. C., Mortenson, L. E., Carnaham, J. E.: J. Biol. Chem. **238**, 1141 (1963).
443. Valentine, R. C.: Bacteriol. Rev. **28**, 497 (1964).
444. Van Tieghem, P. E. L.: Bull. Soc. Bot. France **24**, 128 (1877).
445. Vetter, H., Knappe, J.: Hoppe-Seylors Z. Physiol. Chem. **352**, 433 (1971).
446. Vinayakumar, M., Kessler, E.: Arch. Microbiol. **103**, 13 (1975).
447. Wall, J. D., Weaver, P. F., Gest, H.: Nature (London) **258**, 630 (1975).
448. Waschsman, J. T., Barker, H. A.: J. Biol. Chem. **217**, 695 (1955).
449. Warburg, O.: Heavy metal prosthetic groups and enzyme action, Chapter XVIII. Oxford University Press, Oxford 1949.
450. Waring, H. S., Werkman, C. H.: Arch. Biochem. **4**, 75 (1944).
450a. Weiner, P. J., Zeikus, J. G.: Appl. Environ. Microbiol. **33**, 289 (1977).
451. Weinberg, M., Nativelle, R., Prevot, A. R.: In "Les Microbes Anaerobies" (Ed. Masson et Cie), Monographie de l'institut Pasteur, Paris 1937, pp. 1186.
452. Weissman, J. C., Benemann, J. R.: Appl. Environ. Microbiol. **33**, 123 (1977).
453. Weller, I., Doemel, W., Briik, T. D.: Arch. Mikrobiol. **104**, 7 (1975).

454. Westlake, D. W. S., Wilson, P. W.: Can. J. Microbiol. **5**, 617 (1959).
455. White, D. C., Sinclair, P. R.: Advan. Microbial. Physiol. **5**, 173 (1971).
456. White, A., Handler, P., Smith, E. L.: In "Principles of Biochemistry" (4th edition) McGraw-Hill, New York 1973, p. 357.
457. Whiteley, R. H., Douglas, H. C.: J. Bacteriol. **61**, 605 (1951).
458. Whiteley, H. R.: J. Bacteriol. **63**, 163 (1952).
459. Whiteley, H. R., Ordal, E. J.: J. Bacteriol. **70**, 608 (1955).
460. Whiteley, H. R., Woolfolk, C. A.: Biochem. Biophys. Res. Commun. **9**, 517 (1962).
461. Whiteley, H. R., Pelroy, A.: J. Biol. Chem. **247**, 1911 (1972).
462. Widdel, F., Pfennig, N.: Arch. Microbiol. **112**, 119 (1977).
463. Wieringa, K. T.: Ant. van Leeuwen. J. Microbiol. Serol. **6**, 251 (1940).
464. Williams, J. P., Davidson, J. T., Peck, H. D.: Bacteriol. Proc. **110**, 128 (1964).
465. Wilson, J., Krampitz, L. O., Werkman, C. H.: Biochem. J. **42**, 598 (1948).
466. Wimpenny, J. W. T., Ranlett, M., Gray, C. T.: Biochim. Biophys. Acta **73**, 170 (1963).
467. Wimpenny, J. W. T., Cole, J. A.: Biochim. Biophys. Acta **148**, 233 (1967).
468. Wimpenny, J. W. T.: Biotechnol. Bioeng. **11**, 623 (1969).
469. Winfrey, M. R., Zeikus, J. G.: Appl. Environ. Microbiol. **33**, 275 (1977).
470. Winter, H. C., Burris, R. H.: J. Biol. Chem. **243**, 940 (1968).
471. Wolf, C. G. L.: Brit. J. Exper. Path. **2**, 266 (1921).
472. Wolf, M.: Science **184**, 382 (1974).
473. Wolfe, R. S., O'Kane, D. J.: J. Biol. Chem. **205**, 755 (1953).
474. Wolfe, R. S.: Advan. Microbiol. Physiol. **6**, 107 (1971).
475. Wolin, M. M., Wolin, E. A., Jacobs, N. J.: J. Bacteriol. **81**, 911 (1961).
476. Wolin, E. A., Wolfe, R. S., Wolin, M. J.: J. Bacteriol. **87**, 993 (1964).
476a. Wolin, M. J.: Am. J. Clin. Nutr. **27**, 1320 (1974).
477. Wood, W. A.: In "The Bacteria" (Ed. I. C. Gunsalus, R. Y. Stanier), Academic Press, Inc., New York 1961, p. 59.
478. Wood, N. P., Jungermann, K.: FEBS Lett. **27**, 49 (1972).
479. Woods, D. D.: Biochem. J. **30**, 515 (1936).
480. Woods, D. D., Clifton, C. E.: Biochem. J. **31**, 1774 (1937).
481. Woods, D. D., Clifton, C. E.: Biochem. J. **32**, 345 (1938).
482. Yagi, T., Honya, M., Tamiya, N.: Biochim. Biophys. Acta **153**, 699 (1968).
483. Yagi, T.: J. Biochem. (Tokyo) **66**, 473 (1969).
484. Yagi, T.: J. Biochem. (Tokyo), **68**, 649 (1970).
485. Yagi, T., Maruyama, K.: Biochim. Biophys. Acta **243**, 214 (1971).
486. Yamane, I., Sato, K.: Soil Sci. Plant Nutr. **10**, 127 (1964).
487. Yu, L., Wolin, M. J.: J. Bacteriol. **88**, 51 (1969).
488. Yudkin, J.: Biochem. J. **26**, 1859 (1932).
489. Zajic, J. E., Brosseau, J. D.: First World Hydrogen Energy Conference, 1–3 Mar., 1976, Miami Beach, Florida. Conference Proceedings (Ed. T. Veziroglu), Univ. Miami, Coral Gables, Florida 1976 a, p. 4B–29.
490. Zajic, J. E., Brosseau, J. D.: Paper presented at the 32nd Annual Meeting of the Society for Industrial Microbiology, 14–19 August 1976, Jekyll Island, Georgia 1976b.
491. Zajic, J. E., Brosseau, J. D.: Dev. Ind. Microbiol. **18**, 637 (1976c).
492. Zajic, J. E., Brosseau, J. D., Kosaric, N.: The effect of inorganic sulfur compounds on growth and hydrogen gas production by *Citrobacter intermedius.* Paper presented at the Pacific and Asian Chemical Engineering Conf., 28–31 Aug. 1977. Denver, Colo 1977.
493. Zajic, J. E., Brosseau, J. D.: Hydrogen production using *Citrobacter intermedius.* International Course cum Symp. on Bioconversion of Cellulosic Substances into Energy, Chemicals and Microbial Protein, 7–23 Febr., 1977. New Delhi 1977.
494. Zehnder, A. J. B., Wuhrmann, K.: Science **194**, 1165 (1976).
495. Zeikus, J. G., Weimer, P. J., Nelson, D. R., Daniels, L.: Arch. Micro. **104**, 129 (1975).
496. Zeikus, J. G.: Bacteriol. Rev. in press (1977).
497. Zobell, C. E.: Bull. Amer. Assoc. Pet. Geologists **31**, 1709 (1947).

In Vitro Synthesis of Enzymes. Physiological Aspects of Microbial Enzyme Production

Toshio Enatsu and Atsuhiko Shinmyo
Department of Fermentation Technology
Osaka University, Osaka, Japan

Contents

The present status of *in vitro* synthesis of enzymes is discussed from the viewpoint of the physiology of microbial enzyme production. Physiological features of several microbial productions of excreted hydrolases are surveyed. The kinetic characteristics of the hydrolytic enzyme productions are accounted for mainly in terms of induction and the stability of the capacity for enzyme synthesis in addition to catabolite repression and anabolite repression. Mechanisms of protein synthesis are outlined with reference to the role of mRNA and ribosomes as well as the role of the sequences of reactions in chain initiation, chain elongation and termination.

In vitro synthesis of specific proteins is surveyed with reference to the criteria governing the *in vitro* product and the role and behavior of the mRNA of eukaryotes and prokaryotes. *In vitro* translation of enzymes is reviewed with reference to the translation of isolated mRNA and the synthesis of animal and microbial secretory enzymes. Interesting features of the cell-free translation of endogenous mRNA are shown. Evidence that the microbial secretory enzyme is synthesized on polysomes bound to membranes in living cells and that mRNA for the synthesis is long-lived and is accumulated is derived from the experimental results of an *in vitro* system.

1. Introduction

1.1 In Vitro Synthesis of Enzymes

Cells of organisms synthesize enzymes, which are special forms of protein capable of catalyzing chemical reactions in organisms. In general, the term "*in vitro* synthesis" has been used synonymously with "synthesis with cell-free preparations" in recent articles. Cells contain a variety of subcellular components: cell wall, cell membrane, cytoplasm and nuclear zone, in prokaryotic cells; cell wall (in plants), cell membrane, cytoplasm, mitochondria, chloroplasts, nucleous, endoplasmic reticula, Golgi bodies and microbodies, in eukaryotic cells. Some cellular components can be successfully isolated in an active state by the disruption of cells and by fractional centrifugation. The study of the cell-free protein synthesis has advantages in elucidating detailed mechanisms and factors required in the steps of protein-forming processes. Biosynthesis of protein occurs on the surface of certain ribosomes by a series of condensations of amino acid residues to form peptide chains. With the participation of a number of essential protein factors and enzymes, and by utilizing the energy of GTP hydrolysis, activated amino acids are added in the order determined by genetic information encoded in the nucleotide base sequence of messenger ribonucleic acid (mRNA). Thus, the genetic information of mRNA transcribed from deoxyribonucleic acid (DNA) is translated, with the commitment of these cellular components into a sequence of amino acids of a polypeptide chain, which determines the ultimate steric structure of the protein molecule. Recent studies show that entire peptide chains of enzymes, possessing catalytic activity, are newly synthesized with exceedingly high fidelity to the message in systems containing these subcellular fractions from the amino acids added to the reaction mixture.
Studies on *in vitro* synthesis of enzymes have been undertaken by many investigators in various fields, following the demonstration in 1961 by Matthaei and Nirenberg of cell-free protein synthesis in a mixture of subcellular fractions of *Escherichia coli* capable of translating mRNA [1, 2]. The study of *in vitro* synthesis of enzymes, rather than general proteins, has the advantage that detection of a particular protein among the peptide chains synthesized in the cell-free system is facilitated by the biological activity of the enzyme protein. Biochemists endeavor to learn the detailed mechanism of the synthesis, and chemical biologists may apply this knowledge in investigating the behavior of organisms. Biochemical engineers may use the information in designing artificial reactor systems for synthesizing protein. In particular, for biochemical engineers involved in microbial enzyme production, the information must allow them to program improved cultivation processes for industrial production. In this article emphasis will be placed on *in vitro* formation of microbial enzymes, as more readers of this book series may be concerned with microbial enzymes to some extent.
Remarkable progress in protein synthesis has been made in the past decade and the complexity of the processes involved in the synthesis has been recognized. Evidence for postulated mechanisms of protein synthesis has been accumulating, particularly concerning the steps of polypeptide chain formation: activation of amino acids, chain initiation, chain elongation, chain termination and the ribosome cycle. Information on new features of the stimulating protein factors is also becoming available. The role of

messenger RNA, which transmits genetic information to the protein, is well understood both in prokaryotes and eukaryotes. However, the behavior of the molecule in the protein synthesis system of a particular species of protein is obscure in all but a few cases. The extraordinarily high stability of mRNA in the living cell system has also been indicated in some cases. In this article, the review will be mainly restricted to aspects of microorganisms and mRNA.

1.2 Messenger RNAs in Microbial Enzyme Production

Before considering *in vitro* synthesis of enzymes, it is appropriate to give the reader some physiological insight into the *in vivo* synthesis of microbial enzymes. Kinetic analysis of the mass production culture of microbial enzymes is an essential approach in improving the productivity of the culture after optimizing the conditions for production such as medium composition, aerating rate, and agitation speed. An understanding of the physiological factors underlying the production pattern of a given culture is important in determining the course of the culture operation for production. The importance of a physiologically significant approach has been stressed by Terui and his group [3–7] at Osaka University. By relating the rate of production to the growth rate of the microorganism, a general classification of the process patterns of product formation has been proposed [8]. A variety of process patterns of enzyme production are reasonably well accounted for by the individual physiological factors. The production patterns of some microbial hydrolases are especially well explained by kinetic models in which the quantitative relationship of the change in enzyme-forming activity with the specific growth rate and the longevity of the enzyme-forming system is basically described by consideration of the physiological factors such as growth limiting nutrient, type of inducer, catabolite repression and anabolite repression. Examples of production include glucamylase [5], polygalacturonase [9], acid protease [10] of *Aspergillus niger*, α-amylase of *Bacillus amyloliquefaciens* (formerly classified as *B. subtilis*) [4, 6], cellulase of *Penicillium variabile* [11] and tannase of *A. oryzae* [12]. The production patterns of these secretory enzymes are classified into three types (Fig. 1).

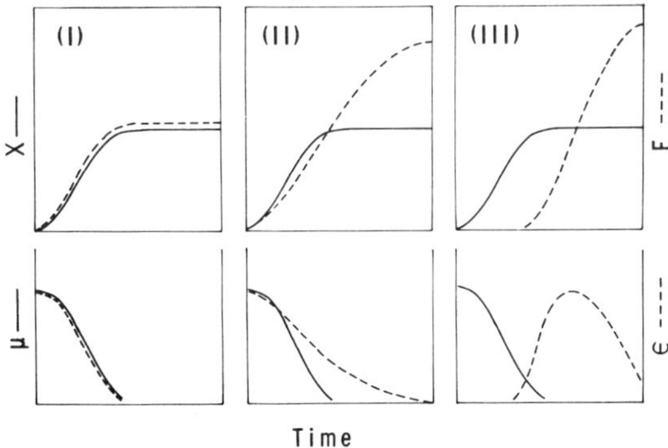

Fig. 1. Schematic process patterns of the production of hydrolases. X: cell concentration, E: enzyme concentration, μ: specific rate of growth, ϵ: specific rate of enzyme formation.
Type I: growth-associated production pattern. Type II: modified pattern of growth-associated production in which production continues after growth has ceased. Type III: growth-dissociated production pattern

In general, the limiting factor for the activity of the enzyme-forming system is the stability of the mRNA for the specific enzyme produced. The mRNA for extracellular enzymes is extraordinarily stable, having a half-life of 5–8 h at 30–35 °C (Table 1). In contrast, mRNAs for intracellular enzymes are shown to be very labile, as indicated by Jacob and Monod [13–17]. In some studies the unusually high stability has tentatively been attributed to the special state of existence of the mRNA in the cells [18, 19].

Table 1. Half-lives of messenger RNAs for secretory hydrolases as well as for intracellular enzymes of microorganisms

Enzyme	Microorganism	Temp (°C)	Half-life of mRNA	Ref.
Secretory hydrolase				
α-Amylase	*B. amyloliquefaciens*	35	6.3 h	[4]
		30	7.7 h	[5]
Glucamylase	*A. niger*	35	6.5 h	[5]
		30	8.6 h	
Acid protease	*A. niger*	30	6.9 h	[7]
		25	13.8 h	
Polygalacturonase	*A. niger*	30	6.9 h	[9]
Lipoprotein of outer membrane[a]	*E. coli*	37	11 min	[39]
Intracellular enzyme				
β-Galactosidase	*E. coli*	37	1 min	[14]
		37	1.3 min	[15]
		30	1.7 min	
Arginase	*A. nidulans*	30	2.7 min	[16]
Kynureninase	*Neurospora crassa*	25	15 min[b]	[17]
		20	32 min	
DHBA[c] carboxylyase	*Trichosporon cutaneum*	30	8 min[b]	[17]
		20	23 min[b]	
Catechol oxygenase	*T. cutaneum*	20	3 min[b]	[17]

[a] Shown for comparison.
[b] Half-lives of off-coding mRNA estimated from experiments in which cells which had been induced and then deinduced in the presence of cycloheximide, were allowed to develop the capacity for the enzyme in the absence of the inhibitor, while the capacity as a function of time was observed.
[c] 2,3-dihydroxybenzoate.

A typical example of a Type I production pattern is seen in tannase production by *Aspergillus oryzae* var. *pseudoflavus* grown on tannin or gallic acid as the sole carbon source [12]. Almost equal amounts of the enzyme accumulate intracellularly and extracellularly with growth of the fungus and the production ceases with the end of growth caused by exhaustion of the carbon source (Fig. 2). Enzyme production also ceases upon removal of inducing substrate, this fact suggests that the mRNA for the tannase is very unstable. Polygalacturonase is induced by galacturonate or pectin in the culture of *A. niger* U20-2-5, an adenineless strain [9]. When galacturonate is used as an inducer, the

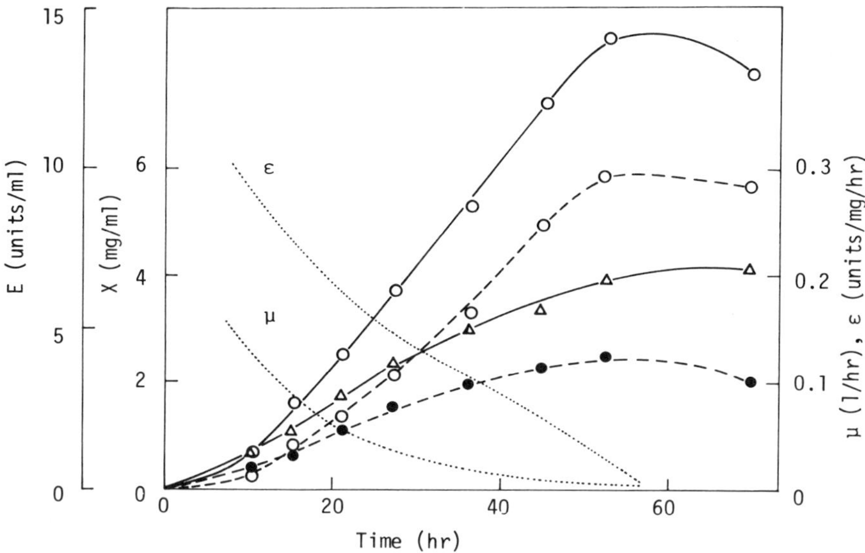

Fig. 2. Time course of tannase production in gallate medium by *A. oryzae.* Mycelia grown in a glucose medium were transferred to and incubated in a gallate (2%) medium. The cell mass, X (△); tannase activities in filtrate (---○---) and in cell extract (---●---) and total tannase activity, E (—○—), were observed. μ: specific rate of growth, ϵ: specific rate of enzyme formation

Type II pattern of production is exhibited, while a typical profile of Type III is observed with pectin as the inducer (Fig. 3). The difference between them is attributed to the difference in the induction lag in the cultures, which results from the repression caused by glucose contaminating the pectin preparation. In *A. niger* U20-2-5, syntheses of

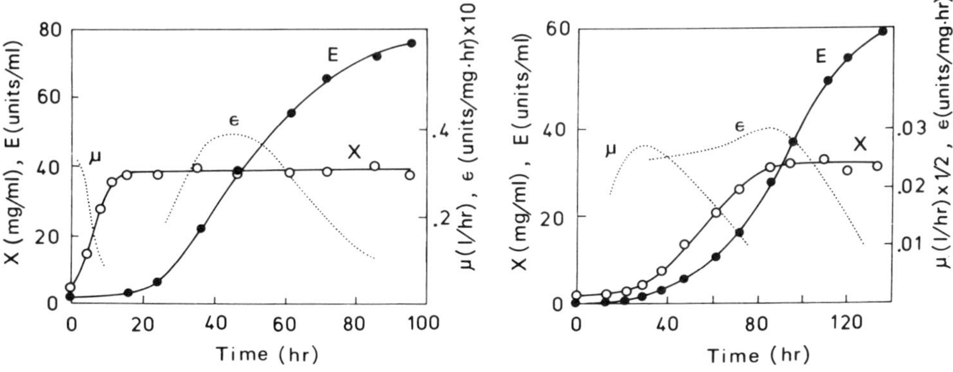

Fig. 3. Time course of polygalacturonase production by *A. niger* U20-2-5. Mycelia were cultured in a pectin medium (figure on the left) and in a galacturonate medium (figure on the right). X: cell concentration, E: enzyme concentration, μ: specific rate of growth, ϵ: specific rate of enzyme formation

acid protease and glucamylase are induced by peptone and starch (or maltose), respectively, and dissociated with the growth exhibiting the Type III pattern [5, 7] (Figs. 4 and 5). The negative correlation of the specific production rate with a specific growth rate was originally ascribed to hypothetical anabolite repression (formerly growth-associated repression), which is directly correlated with anabolism and results from unknown reaction(s) [3]. In the case of acid protease, since the production pattern was not altered by the presence of glucose and citrate in the medium, catabolite repression could not account for the mechanism of anabolite repression.

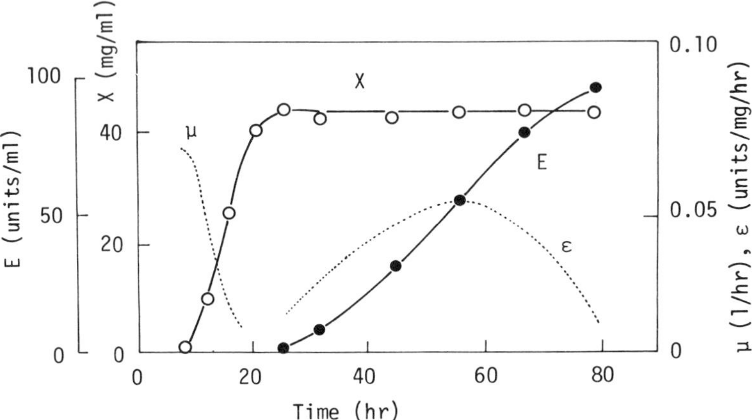

Fig. 4. Time course of acid protease production by *A. niger* U20-2-5. X: cell concentration, E: enzyme concentration, μ: specific rate of growth, ϵ: specific rate of enzyme formation

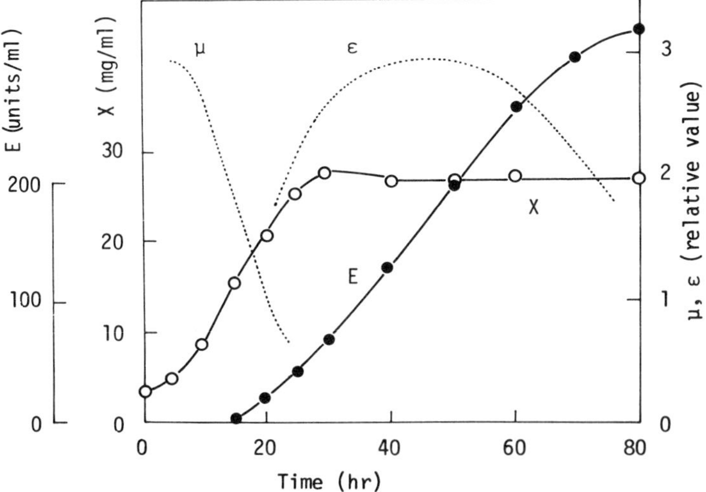

Fig. 5. Time course of glucamylase production by *A. niger* U20-2-5. X: cell concentration, E: enzyme concentration, μ: specific rate of growth, ϵ: specific rate of enzyme formation

When uninduced cells in the stationary phase, caused by limitation of the essential nutrient, adenine, were induced with peptone, the addition of adenine resulted in the repression of the enzyme synthesis along with the resumption of growth [20]. Furthermore, the production pattern of the enzyme by a double mutant requiring adenine and methionine, which was derived from U20-2-5, was not altered by a change of the growth limiting nutrient from adenine to methionine. From these observations, it is concluded that the repression of the production during growth is caused by neither intermediate(s) in adenine-related metabolism nor by a lag of induction, implying strongly that the repression is correlated with anabolism *per se.*

The enzyme production by mycelia during the stationary phase caused by adenine starvation was attributed to *de novo* synthesis of mRNA of high stability, which occurs as part of the metabolic turnover of RNA in living cells. The turnover of RNA was confirmed by incorporation of a labeled nucleotide base into the cell RNA fraction and by the inhibition of nucleic acid synthesis with certain inhibitors. These observations were supported by a deinduction experiment which revealed the longevity of the messenger RNA for the enzyme [20].

The system of polygalacturonase production exhibited a novel behavior for the enzyme synthesis toward catabolite [21]. The synthesis by the mold strain U20-2-5 of polygalacturonase and protease is suppressed by the addition of sugar substances, including glucose, but not by intermediates in the TCA cycle. There is some difference in the effect of glucose on the production of the two enzymes. While the synthesis of the polygalacturonase is inhibited rapidly and completely, that of the acid protease is suppressed moderately after a lag (as in Fig. 6). These facts indicate that there exists some difference in the level of repression. In fact, the addition of glucose in the presence of

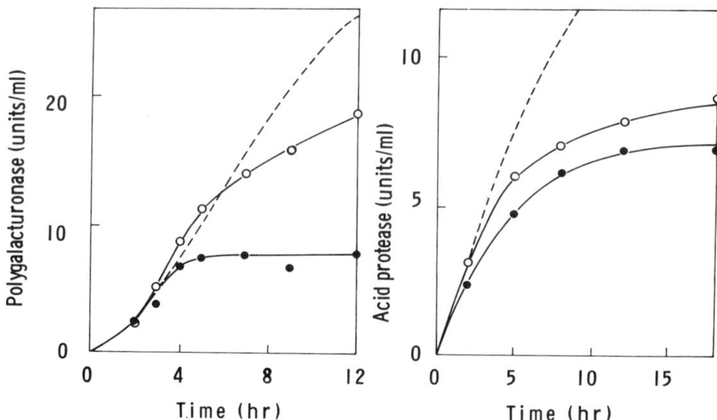

Fig. 6. Effect of glucose on the production of polygalacturonase and acid protease in cultures with added actinomycin S_3. The enzyme activities were observed in the culture with added actinomycin S_3 (–○–), with actinomycin S_3 plus glucose (2%) (–●–) and without addition (-----) as a control measurement. Actinomycin was added to polygalacturonase and acid protease cultures at 2 h and 0 h, respectively

actinomycin revealed that repression of the polygalacturonase synthesis occurred at the level of translation (Fig. 6). In contrast, the lowered production of the acid protease was attributed to the interference with transcription due to the fact that glucose did not antagonize the uptake of peptone but did inhibit induction by it in non-growing mycelia. It was also revealed that the messenger capacity for polygalacturonase synthesis was preserved during the inhibition of translation caused by cycloheximide, while the capacity for the enzyme synthesis did not accumulate in the presence of glucose in a medium containing an inducer (Fig. 7). These observations indicate that glucose suppresses not only translation but also transcription of the messenger for polygalacturonase.

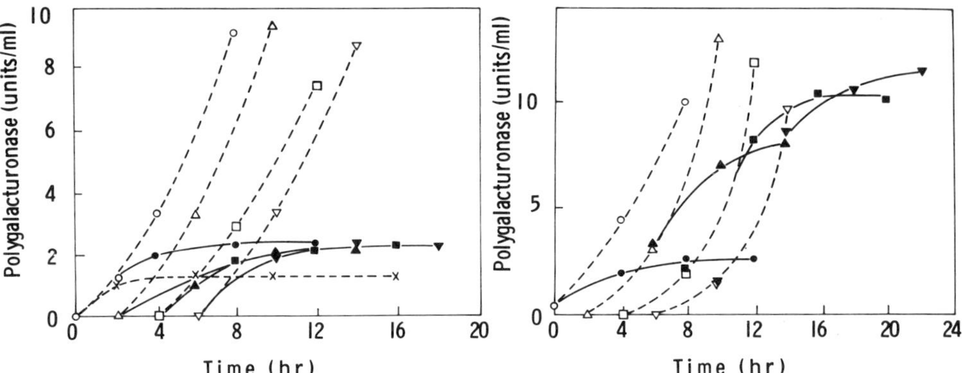

Fig. 7. Accumulation of polygalacturonase-forming capacity during inhibition at translational level. Translation in mycelia in an induced culture was inhibited by the addition of glucose (left) and cycloheximide (right) at 0 h (○), 2 h (△), 4 h (□) and 6 h ( ). Thereafter the mycelia were transferred to an inhibitor-free medium and development of the activity was observed in the presence (solid symbols) and absence (open symbols) of actinomycin S_3. In a control culture, mycelia continued to incubate in the same inhibitor medium (--x--)

The facts described above indicate that the production of microbial hydrolase might have a close correlation to the capacity for enzyme synthesis, which represents the level of mRNA, and that in general the stability of mRNA for the enzyme is exceedingly high. It is natural that special attention is focused on the behavior of mRNA molecules in discussing the *in vitro* synthesis of enzymes.

2. Outline of the Mechanism of Protein Synthesis

The most recent picture of the mechanism of protein synthesis is summarized from recent works which include original papers and reviews [22–30].

2.1 Message and Ribosomes

Genes, units of heredity constituting chromosomes, pass from one generation to the next in a definite manner. The DNA of chromosomes contains genetic information and a sequence of DNA molecules is responsible for the synthesis of a single protein chain. DNA has a structure in which two long polynucleotide chains are wound around each other to form a double helix. The two chains are oriented in opposite directions and held together by hydrogen bonding between complementary base pairs, adenine-thymine and guanine-cytosine. DNA-dependent RNA polymerase catalyzes the formation of a polymer of RNA from ribonucleoside 5′-triphosphates, which is complementary to the DNA template.

In the initial step of protein synthesis, 20 different constituent amino acids are each activated by a specific aminoacyl-tRNA synthetase. The synthetases catalyze the formation of the aminoacyl esters of corresponding tRNAs in a reaction coupled with the pyrophosphate cleavage of ATP.

$$\text{Amino acid + tRNA + ATP} \overset{Mg^{++}}{\rightleftharpoons} \text{aminoacyl-tRNA + AMP + PP}_i.$$

The amino acid transfer to the specific tRNA proceeds in two distinct steps on the enzyme surface. The specificity of the aminoacyl-tRNA (AA-tRNA) synthetase for tRNAs and their corresponding amino acids is very high. Each AA-tRNA, thus formed, is recognized not at the amino acid but at the triplet sequence (anticodon) of the tRNA, which is complementary to the code triplet (codon) on the mRNA. Accordingly, the specificity plays a very important role in translating genetic information to the amino acid sequence. All tRNAs are shown capable of sharing a common confirmation of a 4-branch cloverleaf, despite differences in their base sequences. The anticodon triplet is located on the loop which is thought to be tertiarily most distal from the 3′ end binding (to be attached) to amino acid. Ribosomes are regarded as the field of protein synthesis which occurs on the surface of the particles under the direction of the triplet code embodied in the mRNA. Bacteria and other prokaryotic cells contain 70 S ribosomes which dissociate reversibly into 50 S and 30 S subunits. These ribosomes are composed of a few RNA chains and about 50 polypeptide chains, which can be differentially extracted with salt solution. In addition to the 70 S ribosomes in the mitochondria (and chloroplasts), eukaryotic cells possess 80 S ribosomes in their cytoplasm, which are dissociated into 60 S and 40 S subunits and are similar in their constitution of RNA and protein to prokaryotic ones.

2.2 Chain Initiation

Synthesis of a single protein peptide chain takes place exclusively on the surface of ribosomes as the result of their movement along the mRNA in the direction from 5′ to 3′ with concomitant sequential addition of amino acid residues from AA-tRNAs. In the prokaryotic system, the peptide chain starts with methionine, which corresponds to the initiation codon AUG. The first methionine residue is introduced as N-formylmethionyl-tRNA (fMet-$tRNA_F$), which is formed by formylation of methionyl-$tRNA_F$ with a specific enzyme. The initiating methionyl-$tRNA_F$ is quite different from methionyl-

$tRNA_{Met}$, the species which functions as the donor of methionine residue inserted into the peptide chain.

A portion of the ribosomes exists as 50 S and 30 S subunits as the result of the dissociation of 70 S ribosomes after the completion of protein synthesis. At first, messenger RNA and initiating tRNA do not bind to the 70 S ribosomes but are bound instead to the smaller subunit and form an initiation complex, which then associates with the 50 S subunit. For the formation of the initiation complex three initiation factors (IF-1, IF-2, and IF-3) are found to be required. These are proteins extractable from ribosomes with salt solutions (Fig. 8). IF-3, which is bound to the 30 S subunit, participates in the association with mRNA, while IF-2 possesses affinity for fMet-$tRNA_F$ and GTP and effects the formation of an intermedial complex. IF-1 is required for the formation of the initiation complex composed of the two intermedial complexes. On association of the complex with the 50 S subunit the three initiation factors are released and GTP is hydrolyzed to GDP and inorganic phosphate. The factors seem to have very high specificities which lead to a correct initiation together with the initiating AA-tRNA at the right initiation codon (AUG) on the messenger.

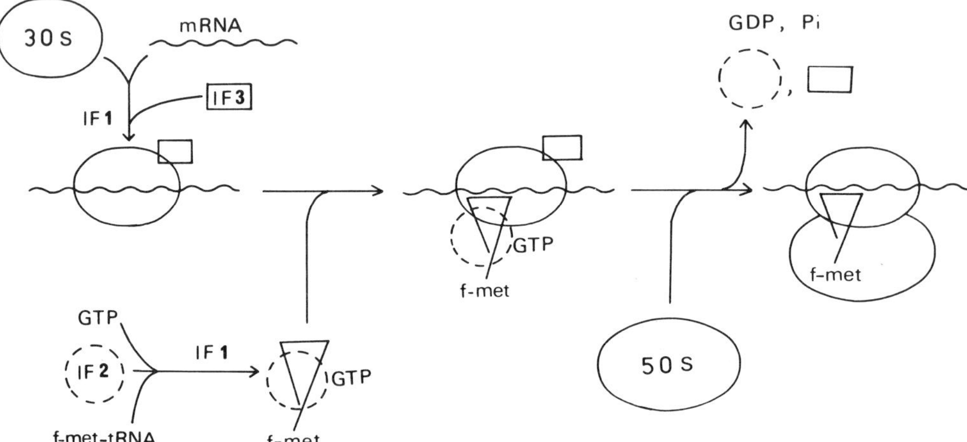

Fig. 8. Scheme of a chain initiation mechanism showing the participation of the mRNA, the 30 S ribosomal subunit, the initiation factors IF 1, IF 2, IF 3, the GTP, the f-met-tRNA and the 50 S ribosomal subunit in the protein synthesis

2.3 Chain Elongation

The process of peptide chain elongation occurs after completion of the functional 70 S ribosome complex and comprises three steps distinguishable by the requirement of specific protein factors:

1) binding of AA-tRNA,
2) peptide-bond formation and
3) translocation (Fig. 9).

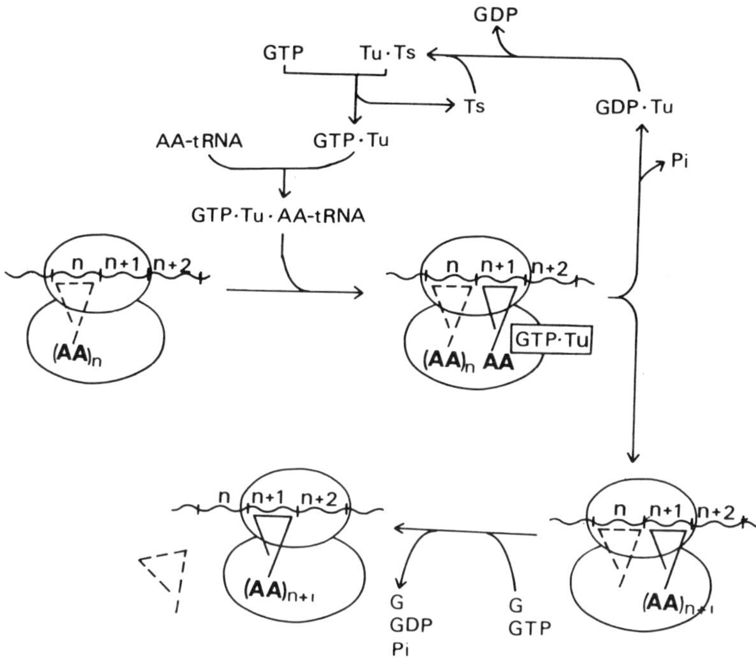

Fig. 9. Schematic presentation of chain elongation in the translation of protein. See text for description

In the binding of AA-tRNA, the incoming AA-tRNA binds itself to the aminoacyl site (site A) of the 70 S ribosome complex. For this binding step to take place, a cytoplasmic protein factor, elongation factor T (EF-T) is shown to be required. The EF-T, which comprises two subunits (Ts and Tu), combines with GTP and forms a complex of GTP and EF-Tu with release of the other subunit, EF-Ts. The complex, GTP-Tu, then associates with the AA-tRNA resulting in an intermedial complex, GTP-Tu-AA-tRNA. The 70 S ribosome complex combines with the tertiary complex which leads the AA-tRNA to the site A with pairing of the anticodon on tRNA to the codon on the mRNA involved in the elongation under way. The GTP bound to the ribosome is immediately hydrolyzed to GDP and inorganic phosphate. Thus the GDP combined with the subunit of the factor is liberated from the ribosome.

In the first peptide-bound formation step the formylmethionyl group of fMet-$tRNA_F$ on the peptidyl donor site (site P) is transferred to the amino group of the AA-tRNA on site A by a specific enzyme, which is a protein component of 50 S subunits. The dipeptidyl-tRNA thus formed and the discharged tRNA for formylmethionine remain bound on the site A and site P, respectively. The processes described above for the first peptide formation is believed to be common to the subsequent formations for peptide linkages of the protein chain.

The peptidyl-tRNA, which is newly formed by the sequential addition of an aminoacyl group at the carboxyl terminal end of the peptide chain, moves from the site A to the adjacent site P on the ribosome surface. The movement is accompanied by a shift of messenger RNA by one codon unit and provides a vacancy at the site A for the incoming AA-tRNA specified by the subsequent codon. This translocation of peptidyl-tRNA requires the presence of both a specific elongation factor G (EF-G) and GTP. GTP is associated with a ribosome after the complex formation with the EF-G. The translocation reaction is driven by energy provided by hydrolysis of GTP to GDP and inorganic phosphate, which are released together with EF-G following the reaction.

In these processes, peptidyl-tRNA at the site A is transferred to the site P and concurrently mRNA is shifted relatively, one codon unite along the surface of ribosome. There must be very complex changes in the peptidyl-tRNA complex and in the enzymes involved in this shift as well as in the ribosomal subunits.

In response to one of three specific termination codons, UAA, UAG and UGA, in mRNA, the ribosome complex completes the peptide chain and releases peptide and mRNA. The last peptide linkage is formed between the carboxyl group of the polypeptide and the amino group of the last AA-tRNA on the site A, the entire polypeptide remaining linked to the tRNA by an ester bond. Three specific protein factors, the release factors R_1, R_2, and R_3, are required for the release of the polypeptidyl-tRNA and the components of the translating complex. The binding of the factor results in a translocation of the polypeptidyl-tRNA from the site A to the site P. The hydrolysis of the ester linkage between tRNA and the peptide occurs by way of the catalysis of peptidyl transferase, which appears to alter the bound release factors. Thereafter, the mRNA and the tRNA are released from the ribosome and the free 70 S ribosomes dissociate into 50 S and 30 S subunits.

In living cells each event in the steps described above takes place sequently under a network of coherent control systems. The efficiency of protein synthesis in living cells is so high that ribosomes start to translate one after the other on a chain of mRNA. Several ribosomes involved in the peptide chain synthesis form a complex cluster, a polyribosome or polysome, which consists of a number of ribosomes, polypeptide chains growing on them and a mRNA strand.

The above described mechanism of protein synthesis has been deduced mainly from observations of the individual reaction steps in the prokaryotic system, mostly bacteria, and is considered common to the eukaryotic system. The main differences in the mechanism occurring in the eukaryotic system are:

1) participation of 80 S ribosomes in the synthesis in the cytoplasm, as shown above,
2) initiation with unformylated methionyl-$tRNA_F$, and
3) the requirement of protein factors which behave slightly differently from those of prokaryotes.

The efficiency of the translation of mRNA is of importance in achieving the *in vitro* synthesis of protein. It may be fairly high in some cases, approaching that of living cells, and so low in other cases that the synthesis can hardly be detected even with a radioactive tracer.

3. In Vitro Synthesis of Particular Proteins

3.1 Identification of the Products of In Vitro Synthesis

The products of the *in vitro* synthesis of protein can be identified by a combination of procedures differing in principle only on the basis of the properties of the protein molecules. Labeling with radioactive amino acids is frequently used to detect very small quantities of protein, synthesized in a cell-free system. The products or their derivatives are fractionated by differences in solubility, by chromatography or by forced migration. Generally, acid precipitation with trichloroacetic acid is used to estimate total protein; the incorporation of amino acid into total protein is assayed making use of radioactivity. Insolubility in organic solvents such as alcohols and acetones, as well as solutions of salts including ammonium sulfate are employed to separate the synthesized native protein species. Chromatographic fractionations are based on differences in the affinity of the protein molecules for the eluting fluid and fixed phase. In gel filtration partition of the molecules of different size occurs when a solution is passed through a column of porous matrices (such as Sephadex and Biogel) possessing pores which allow proteins to enter. Flow through a bed of ion exchangers results in separation of protein molecules according to the degree of electrical charge of the molecules. Matrices of covalently linked molecules having a configuration similar to that of an enzyme substrate and an affinity for specific species of protein can retain species, such as enzymes. The affinity of an antibody, developed immunologically in animals by injection of purified protein as an antigen, can also be used to isolate specific proteins which share a group of three dimensional determinants with the antigen protein. Subsequently, the identity of the spacial determinant of the enzyme synthesized *in vitro* is confirmed by affinity for the substrate or cross-reaction with the antibody. The spacial determinant provides evidence of the identity of the amino acid sequence of that portion of the peptide chain which includes binding sites for the determinant, but not always that of the entire chain.

Forced migration procedures include centrifugation and the application of an electric potential. In the former case protein molecules sediment through a density gradient of a solution such as sucrose. In the latter case, charged molecules are propelled by a force proportional to the charge and inversely proportional to resistance due to the size of the molecule and to the pore size of the sustaining matrix such as polyacrylamide gel. The addition of sodium dodecyl sulfate (SDS) eliminates differences in electric charge by association with the protein and results in a reliable estimate of molecular weight by electrophoresis in the polyacrylamide gel. Fragmentation of the peptide chain is applied in the analysis of protein products fractionated by the above procedures. The specificity of proteolytic attack of enzymes, and/or chemical reagents, allows identification of some or all fragments by comparing the migration of the fragments in two directions. This is the basis of the finger-print technique. Generally, procedures are selected for identification of the product depending on the amount, the biological and chemical properties and purity of the product. Whether the product is newly synthesized (*de novo*) or results from alteration of a preexisting precursor, such as by fragmentation or aggregation, is tested by varying the composition of the synthesizing mixture.

For example, omission of the energy generating system or source of the mRNA results in failure of the *de novo* synthesis.

3.2 Outer Membrane Protein of *Escherichia coli*

A peculiar lipoprotein of the outer membrane of *Escherichia coli* has been shown to be synthesized *in vitro* under the direction of purified mRNA [31], the first biologically active mRNA to be isolated from *E. coli.* The lipoprotein consists of 58 amino acid residues, one unusual amino acid, glycerylcysteine [S-(propane-2′,3′-diol)-3-thio-2-aminopropionic acid] at the amino terminal, two fatty acids bound to the propandiol moiety by ester linkages and one fatty acid attached to the terminal amino group by an amide bond [32–34]. The small lipoprotein is located exclusively in the outer membrane of the cells in both a free form and a bound form which is covalently linked to peptidoglycan [35–38]. The amount of the lipoprotein is estimated to be 7.5×10^5 molecules per cell [37]. The protein is synthesized in free form in the cell and incorporated into the outer membrane without being accumulated in the inner membrane. Studies of the inhibitory effect of antibiotics including rifampicin show that among the biosynthetic capacities for several individual proteins the capacity for the synthesis of the lipoprotein is fairly stable, exhibiting a half life of 11.5 min [39].

The *in vitro* synthesis was undertaken by 1) purifying the mRNA for the lipoprotein, 2) synthesizing protein in a cell-free *E. coli* system and 3) identifying the synthesized protein as the lipoprotein by immunoprecipitation. Starting with 48 g of exponentially growing cells, cell lysate was prepared with lysozyme and sodium dodecyl sulfate. Protein in the lysate was removed by extraction with phenol. RNA was purified by NaCl fractionation, first gel filtration, sucrose gradient centrifugation and second gel filtration. Profiles of the gel chromatography showed that about 85% of the total RNA involved appears at or immediately following the void volume, while 9.5% of the total RNA is recovered as a small peak of samller molecules. The general messenger activity, which is expressed by incorporation of [^{35}S]methionine into acid-precipitable protein in an *E. coli* cell-free system, is associated with the smaller peak of RNA and is much higher than that associated with the larger RNA peak. When the activity of the messenger directing lipoprotein synthesis was assayed by immunoprecipitation of the *in vitro* product with antilipoprotein serum, the smaller RNA fraction exhibited 250 times higher specific activity than the original crude phenol extract. Considering the high recovery of immunoprecipitate, the product directed by the smaller RNA fraction was concluded to consist mainly of the lipoprotein messenger. On slab gel electrophoresis the RNA migrated as two closely adjacent sharp bands, indicating molecular sizes of about 250 and 230 nucleotides.

The *in vitro* product cross-reacts with anit-lipoprotein serum. When the cell-free product was analyzed by sodium dodecyl sulfate-polyacrylamide gel electrophoresis more than 95% of the total radioactivity migrated as a single peak between the internal molecular weight standards cytochrome C and Dansyl-insulin. The identity of the *in vitro* product with the lipoprotein was proved by the agreement of the peptide map of cyanogen bromide-cleaved peptides of the *in vitro* product labeled with [^{14}C]arginine with that of the ninhydrin-stained map of the *in vivo* product of the lipoprotein.

The cells of *E. coli* strain CP78 were labeled with [^{32}P]orthophosphate and total radioactive RNA was prepared by lysing cells with sodium dodecyl sulfate, extracting the lysate and precipitating with ethanol [40]. The specific mRNA for the structural protein in the outer membrane was purified from the total RNA by three successive electrophoreses on polyacrylamide slab gels. The recovery of the radioactive phosphate in the most purified mRNA fraction, which was found to show 90% purity by the finger-print of T_1 ribonuclease digestion, was calculated to be 0.002%. The mRNA was estimated to be 360 ± 10 nucleotides in length. Nucleotide sequence analysis of T_1 oligonucleotide indicated that the mRNA molecule does not contain a polyadenylic acid (poly(A)) sequence at its 3′ end. The fact that the mRNA was not retained on an oligo(dT)-cellulose column supports the above conclusion.

3.3 Preparation and Translation of the Eukaryotic mRNA

Isolation of mRNA for a particular enzyme or protein is a problem of interest in approaching *in vitro* synthesis of enzymes. *In vitro* synthesis of a fragment of rat liver catalase under the direction of fractionated mRNA for catalase prepared by immunoprecipitation techniques has been reported by Uenoyama and Ono [41]. Gel filtration profiles of polysomes from normal rat liver showed that the activity of catalase is eluted together with polyribosomes before a large peak of activity of free soluble catalase. This activity is released by puromycin or ribonuclease treatment and is decayed by the administration of actinomycin D, which is attributed to the nascent chain of catalase. The polyribosomes bearing chains of catalase (or albumin) are precipitated from the polysomal fraction by the immunoprecipitation technique. RNA extracted by the SDS-phenol method from the immunoprecipitates sedimented through a sucrose density gradient as 9.3 S differs from the theoretical size of 16 S for the catalase mRNA. *In vitro* synthesis of catalase using RNA thus obtained was performed in a system containing ribosomes and protein factors. The incorporation into acid precipitate proceeded at a steady rate for 30 min. The incorporation into anti-catalase precipitate corresponded to 46% of the total incorporation. The procedure of antibody specific immunoprecipitation seemed effective for enriching particular mRNA.

In studying the *in vitro* synthesis of protein preparations enriched with biologically active mRNA capable of directing the synthesis of a particular protein have been sought. Methods described so far for preparing eukaryotic mRNA involve many steps for the deproteinization of the RNA by phenol extraction [42–48]. Recently, an improvement for the enrichment of poly(A)-containing mRNA was introduced [42, 43, 47]. Further purification may be performed by sucrose gradient centrifugation [48] or complementary absorption of the poly(A)-chain of mRNA to oligo(dT)-cellulose [43] or poly(U)-Sepharose [49]. Phenol extraction sometimes results in loss of biological activity due to aggregation and fragmentation of the mRNA chain [41, 42, 50]. Improved methods for the purification and assay of eukaryotic mRNA of polysomes were reported by Krystosek *et al.* [51]. This report described how polysome fraction from eukaryotic cells was dissolved in a solution containing 0.5 M NaCl and 0.5% SDS and passed through an oligo(dT)-cellulose column. Elution with a solution of low ionic strength results in fairly purified and biologically active mRNAs with a yield of 1.5–2% of polysomal RNA.

Poly(A) containing mRNA from rabbit reticulocytes, isolated by the oligo(dT)-cellulose method, sediments in a sucrose gradient in three sharp bands at 9 S, 17–18 S, and 28 S. When the messenger activity for the α- and β-globin chain was assayed, the 17–18 S fraction was found to be twice as active as the 9 S fraction per unit amount RNA. The 9 S RNA, when included as sole template, was translated ineffectively, whereas in the presence of 18 S ribosomal RNA the translation was ten times more efficient. Further studies suggested that the 9 S mRNA is an inactive form which is normally activated by the reversible base pairing with specific sites of ribosomal RNA in small subunits and then interacts with the initiation factors and small subunits to form an initiation complex [52].

Translational capacities of rabbit and duck reticulocyte cell-free systems were compared using mouse hemoglobin mRNA which is heterologous to the system [53]. The products, α- and β-globin chains, are estimated using chromatography. In a rabbit reticulocyte cell-free system the mouse mRNA directs synthesis of a significant amount of the β-chain, while the same mRNA resulted in the synthesis of both α-globin and β-globin chains in a duck system. Endogenous mRNA may be replaced with exogenous mRNA, but the endogenous mRNA is translated with 3 to 4 fold higher efficiency than the exogenous mRNA in both cell-free systems.

In studying the translation of the ovalbumin mRNA from hen, a reticulocyte lysate system has been used [54]. Ovalbumin mRNA was translated similarly to endogenous globin mRNA in the system. Hemin prolonged the ovalbumin translation as is observed in globin synthesis, presumably by preventing formation of an inhibitor of polypeptide chain initiation [55–60]. The rate of elongation on ovalbumin mRNA, estimated by measuring mRNA transit time described by Fan and Penman [61], was 0.86 amino acids per second at 26 °C, nearly identical to that of globin translation [62]. Sedimentation analysis of polysomes indicated that ovalbumin is synthesized after 10 min incubation on a polysome composed of 9 ribosomes. Ovalbumin mRNA is not degraded in the system. The size of the polysome decreases gradually to 3 ribosomes during incubation after 70 min. The efficiency of the translation of albumin mRNA was estimated by two independent methods to be 18 and 45 molecules translated per single mRNA molecule. This is several folds less than that of globin mRNA [54].

When heme is deficient protein synthesis in reticulocyte lysate is markedly suppressed [54, 55, 63]. In such a case, there is rapid formation or activation of an inhibitor which arrests polypeptide chain initiation [64–69]. The development of inhibitory activity does not require protein synthesis [64]. Using a purified inhibitor, Ranu *et al.* [70] studied the relationship between the inhibition and initiation factor, IF-MP, which binds to Met-$tRNA_F$ in the eukaryotic system and corresponds to the prokaryotic factor IF-2. Inhibition both by the inhibitor and due to heme-deficiency resulted in dissociation of polysomes and an accompanying increase in the 80 S ribosomes. The inhibition is reversed by IF-MP in both cases. Results of additional study indicated that one of the components inactivated by an isolated inhibitor may be the initiation factor IF-MP. The translational inhibitor from heme-deficient reticulocyte is associated with a protein kinase [71]. When chromatographed on a phosphocellulose column, the inhibitor preparation gave two distinct fractions exhibiting protein kinase [72]. One did not inhibit the protein synthesis, whereas the active inhibitor fraction phosphorylated the small subunit of IF-MP, migrating as a 38000 dalton band, but did not phosphorylate either the large subunits of the IF-MP nor the 40 S ribosomal subunits.

Similar protein kinase activity associated with the hemin-controlled inhibitor was reported by Kramer *et al.* [73]. The purified preparations of the translational inhibitor contained cAMP-independent protein kinase activity which also phosphorylates a smaller subunit of the IF-MP. An antibody against the purified inhibitor obtained from goat serum neutralizes the inhibitory activity as well as the protein kinase activity. These findings support the hypothesis that the translational inhibitor is a cAMP-independent protein kinase possessing the capacity to phosphorylate a subunit of the initiation factor.

Studies described above on the translation of a purified preparation of mRNA for particular proteins in the eukaryotic cell-free system have elucidated the detailed features of the parameters of translation thus allowing some comparison of *in vitro* synthesis with *in vivo* synthesis.

4. In Vitro Synthesis of Enzymes

4.1 Secretory Enzymes of Animals

In vitro synthesis of pigeon pancreatic amylase in relation to the transport of secretory protein has been studied by Redman *et al.* [74]. Generally, enzymes and proenzymes for export are synthesized in exocrine cells. The secretory mechanism postulated [75] involves:

1) vectorial transfer of the newly synthesized protein from ribosomes through endoplasmic reticulal membrane to cisternae and

2) further transport through the cavities of the endoplasmic reticula to the Golgi complex, where the enzyme proteins are concentrated.

They showed unequivocally that the pancreatic amylase is synthesized *in vitro* on ribosomes bound to membranes of microsomes, which correspond to endoplasmic reticula, and is eventually transported to the deoxycholate-soluble microsomal fraction, which corresponds to the content of microsomal vesicle, the cisternal space.

The microsomal system was obtained from a post-mitochondrial supernatant prepared by fractional centrifugation of fresh pancreas homogenate. Microsomal pellet and S-105 supernatant fluid were obtained by 105000 × g centrifugation for 90 min. The microsomal pellet contained microsomal vesicles which originate from endoplasmic reticula in the intact tissue and consist of polyribosomes and membrane vesicles. The *in vitro* system for protein synthesis, consisting of the microsomal fraction, S-105 supernatant, ATP-generating mixture, metallic ions and nucleoside triphosphates, incorporated radioactive amino acid into acid-insoluble protein as well as into a fraction absorbed specifically to glycogen in which amylase is exclusively enriched. The radioactive amylase preparation was found to be homogeneous judging from coincident sharp peaks for amylase activity, protein and radioactivity on the chromatogram from a DEAE-cellulose column.

Electrophoretic study of the peptide obtained by trypsin digestion of the fraction of radioactive amylase showed that the spot patterns of amylase synthesized *in vivo* and *in vitro* are very similar. This fact indicates that the amino acids are incorporated into the amylase chain over its entire length rather than by partial incorporation. The radioactive amylase first appears bound to the ribosomes and is then transferred to the deoxycholate-soluble microsomal subfraction. The deoxycholate-soluble fraction corresponds mostly to the content of the microsomal vesicles. When radioactive amino acids are incubated with the *in vitro* synthesis system about 20% of the amylase formed appears as soluble enzyme in the incubation system and the rest remains bound to microsomes and polyribosomes. The pancreatic microsomal system is thought to be capable of transporting a large proportion of the newly synthesized amylase across the endoplasmic reticulal membrane into the cavity space. Thus the mechanism of excretion of extracellular enzymes in higher organisms is supported by a group of reports [76, 77] including the above study.

In contrast to secretory enzymes, a non-secretory enzyme, arginase, was observed to be synthesized preferentially on free polysomes of rat liver in an *in vitro* system [78]. Support for the hypothesis that secretory protein is synthesized exclusively on polyribosomes bound to membranes of the endoplasmic reticulum seems to be increasing. Arginase was chosen because it exists without being discharged by lever cells only in the supernatant of liver extract and because information on the purified enzyme is available. Both free and membrane-bound polyribosomes were isolated from rat liver homogenate by fractional centrifugation. The supernatant containing soluble components was passed through a Sephadex column to decrease the level of low molecular substances such as amino acid. Free and membrane-bound polysomes were incubated with [^{3}H]leucine and [^{14}C]leucine, respectively, in a cell-free protein synthesizing system containing sephadex-treated cell sap. The two labeled supernatant fractions of the reaction mixture were combined. After nonspecific proteins were removed, the arginase fraction was prepared by an immunoprecipitation technique using antibody against rat liver arginase. The ratio of radioactivity of the arginase fraction to that of the total protein fraction represents the activity of arginase synthesis. The ratio for the free polyribosome system was 5 to 10 times that of the membrane-bound polyribosomal system. The results indicate that arginase in liver cells is preferentially synthesized by free polyribosomes.

4.2 Enzymes of *E. coli*

The *in vitro* synthesis of tryptophanase [79] is shown by polyribosomes isolated from the cells of *E. coli* which are induced for tryptophanase. These polyribosomes possess a small amount of tryptophanase activity. When the induced polyribosomes are incubated with the enzyme synthesis system, the increase in tryptophanase activity associated with polyribosomes is time-dependent and is inhibited completely by chloramphenicol. [^{14}C]leucine incorporation into the protein, which migrates as tryptophanase on a DEAE-cellulose column, was observed to increase concomitantly with the net increase of tryptophanase activity after 45 min incubation of the system. The fractions of the radioactive product which had enzyme activity and were chromatographed on the DEAE-cellulose column behaved identically with tryptophanase on electrophoresis on polyacrylamide gel as well as on centrifugation in a sucrose gradient.

When the distribution of radioactive leucine in the carboxyl region of the newly labeled tryptophanase was examined by digesting the protein with carboxypeptidase, the radioactivity was shown to be more highly concentrated towards the carboxyl terminus with shorter incubation times for protein synthesis. This indicates that peptides synthesized for a shorter incubation time contain labeled amino acids in shorter sequences at the terminal of the tryptophanase chain. These observations present evidence for the synthesis of tryptophanase molecules *in vitro.*

4.3 Lysozyme of T_4 Bacteriophage

In vitro synthesis of enzymatically active lysozyme directed by bacteriophage T_4 RNA was first achieved by Salser *et al.* in 1967 [80]. At that time there had been only a few studies on *in vitro* synthesis of specific proteins, such as hemoglobin and RNA virus proteins, none of which showed enzyme activity. Because of the advantages of achieving the *in vitro* synthesis of active enzymes in a well characterized system, they focused on DNA-dependent synthesis such as those of β-galactosidase [81] and tryptophan synthetase [82]. They attempted to establish cell-free synthesis using the T_4 RNA fraction and the extracts of uninfected cells of *E. coli.* This system has the advantage that the lysozyme assay is quite sensitive and the enzyme fairly well characterized showing molecular weight of 18000 dalton [83, 84]. T_4 infection of the cells results in a remarkable increase in the amount of T_4 mRNA while the use of the RNA excludes possible artifacts, chain completion or activation of pre-existing enzyme.

The RNA for directing *in vitro* synthesis of lysozome was prepared by infecting the cells of an RNase-deficient mutant of *E. coli,* RNase I_{10}^-, with T_4 phage and incubating the infected cells for 20 min. After lysing the cells with lysozyme, the phage RNA was extracted with phenol in the presence of 1% SDS.

The crude extracts from cells of *E. coli* strain RNase I_{10}^- were prepared by grinding with alumina or bursting in a French press as described by Nirenberg and Matthaei [1, 2]. The cell-free S-30 extracts were obtained by centrifugation at 30000 $\times$ g followed by dialysis against a Tris buffer containing magnesium acetate, KCl and mercaptoethanol. The incubation mixture contained S-30 extract, 70 S ribosomes, [^{3}H]leucine, ATP, GTP, polyethylene glycol (as stimulator of the synthesis), an energy-generating mixture (phosphoenol pyruvate and pyruvate kinase) and RNA.

The protein synthesis was estimated by measuring the incorporation of [^{3}H]leucine into acid insoluble material, which increased with increasing RNA. The maximum of the protein synthesis with T_{17} RNA used for comparison occurred at a concentration of 0.45 mg/ml and was 2.5 times higher than that with T_4 RNA from the cells infected for 20 min (T_4 20 min RNA) at a RNA concentration of 1.6 mg/ml, the highest concentration tested. The RNAs extracted from uninfected *E. coli* and from cells infected for

5 min (T_4 5 min RNA) were 2.3 and 1.8 times less active, respectively, than T_4 20 min RNA on a weight basis.

In a lysozyme assay using dead cells as substrate, only the preparation obtained with T_4 20 min RNA showed the enzyme activity. Chloramphenicol and puromycin coordinately inhibited lysozyme synthesis and amino acid incorporation. Treatment with T_1 RNase eliminated the capacity to direct protein synthesis.

The time courses of amino acid incorporation and the development of enzyme activity in the system were similar with the exception of a relative decrease in lysozyme synthesis in the later part of the reaction. When 0.4 mg/ml RNA was used, lysozyme activity reached a maximum after 30 min incubation.

The optimal concentration (7.5 mM) of magnesium was the same for incorporation and lysozyme synthesis, while at concentrations greater than optimal relatively less lysozyme was synthesized. An estimation based on data and reasonable assumption indicates that 4% of total protein formed *in vitro* corresponds to lysozyme. This value is similar to the fraction of late phage specific protein synthesis which is actively forming lysozyme in the cells. These results suggest that the development of lysozyme activity in the *in vitro* system exhibits the following features of *de novo* synthesis:

1) The synthesis is dependent on specific mRNA and inhibited by inhibitors of protein synthesis and RNase,

2) the appearance of lysozyme is similar to the incorporation of amino acids in kinetics and the dependence on magnesium.

Chromatography on Amberlite IRC 50 resin leads to high purification of the enzyme with 6% recovery. Chromatography of lysozyme labeled with different isotopes exhibited almost complete identity for the *in vitro* and the *in vivo* products.

Characterization of the lysozyme messenger by the same group revealed that the active lysozyme messenger appears in a sharp peak just ahead of the 16 S rRNA in a sucrose gradient fractionation and the minimum size of the messenger coding for lysozyme containing 164 amino acids is about 500 nucleotides, which corresponds to a sedimentation rate of 9.3 S [85]. Heat-treatment of the sucrose-fractionated messenger at 80 °C 12 min resulted in a 25% increase in the lysozyme-directing activity in the 17–18 S peak and a concomitant disappearance of the shoulder at 30 S. Column chromatography on benzoylated DEAE-cellulose gave 10-fold purification and suggested aggregation with rRNA.

In 1970 Wilhelm and Haselkorn estimated the rate of chain growth of T_4 lysozyme *in vitro* [86]. Both aurintricarboxylic acid (ATA) and poly(AUG) are known to inhibit initiation of protein synthesis by binding to 30 S subunits and preventing the attachment of mRNA. This kind of inhibitor, when added to a system where protein synthesis is underway, should interfere with new initiation events but permit normal elongation of chains already begun. The increment of enzyme at sufficient time after addition of the inhibitor should correspond to the number of peptide chains completed by active ribosomes. By estimating the time required to synthesize the amount of enzyme corresponding to the increment in the absence of an inhibitor during a period of steady formation in the time course of protein synthesis, the rate of chain growth can be derived. In their system, lysozyme synthesis was observed to be linear for at least 9 min at 25 °C. By tracing the capacity for lysozome synthesis after addition of ATA at various times after initiation of synthesis, the completion times for lysozyme and radioactive protein synthesized *in vitro* were estimated to be 1.75 min and 1.5 min, respectively, at 25 °C. When poly(AUG) was used, the estimated time for completion of lysozyme synthesis was 1.55 min, which agrees well with that observed with ATA at 25 °C. When the *in vitro* system was incubated at 30 °C the completion time for lysozyme synthesis was calculated to be 0.9 min. The chain elongation rate of lysozyme peptide can be derived from the completion time, assuming that the enzyme activity of lysozyme is fully expressed immediately after completion of the peptide chain. The estimated rate of chain elongation is 1.6 amino acids per sec at 31 °C. If the temperature dependence of the elongation rate for lysozyme at 25 °C and 31 °C is applied to estimate the rate at 37 °C *in vitro,* 6 amino acids per sec would be obtained. The elongation rate for β-galactosidase in uninfected *E. coli* cells at 37 °C is estimated to be 15 amino acids per sec [87, 88]. The T_4 RNA-directed system for *in vitro* synthesis is very inefficient with regard to the yield of protein per mRNA molecule. The low efficiency is attributed to poor initiation. Lysozyme messenger activity is associated with RNA, exhibiting a sedimentation coefficient of about 17 S.

In addition to T_4 RNA-dependent *in vitro* synthesis, papers have also appeared on DNA-dependent *in vitro* synthesis of T_4 phage enzyme [89, 90]. α-Glucosyl and β-glucosyl transferases develop in the *E. coli* cells, specifically after T_4 bacteriophage infection. The assay for DNA glucosylation has been performed with high sensitivity in a system containing UDP-[^{14}C]glucose and DNA of a T_4 phage mutant, deficient in transglucosylase.

When T_4 DNA, which is prepared by mild extraction from mature phage, is included in the incubation mixture for *in vitro* protein synthesis containing ribosomes and a S-100 protein fraction from uninfected *E. coli* (K12 AB) cells incorporation of [^{14}C]leucine into hot acid-insoluble protein is stimulated 3 to 4 fold, compared to an endogenous incorporation after 30 min incubation [89]. The incorporation of radioactive amino acids increases with an increase in the concentration of T_4 DNA added to the system, while the enzyme synthesis is nearly saturated by a DNA concentration of 50 μg/ml. The β-glucosyl transferase synthesis is prevented by inhibitors of protein synthesis as well as actinomycin D and by the omission of the S-100 fraction, ribosomes or tRNA. The kinetics of [^{14}C]leucine incorporation and glucosyl transferase synthesis observed in a large incubation mixture revealed that although leucine incorporation was linear for 40 min, the enzyme activity appeared after a lag of 10 min. The enzyme activity continued to increase for 20 min but showed a continuous decrease starting at 30 min, suggesting hydrolysis of protein. The amount of total protein and β-glucosyl transferase synthesized under direction of T_4 DNA in a reaction mixture of 0.025 ml are calculated to be 0.6 μg and 0.18 μg, respectively, on the basis of the specific activity of the purified enzyme and the leucine content in total protein. This corresponds to the synthesis in 0.05 ml reaction mixture of as much β-glucosyl transferase as is found in 0.3 to 1.1×10^7 cells infected for 15 min. The DNA-dependent *in vitro* synthesis of the lysozyme of bacteriophage T_4 [90] has been demonstrated by the same authors using the same *in vitro* system as that for the β-glucosyl transferase synthesis [89]. The radioactive amino acid incorporation into protein was stimulated 6-fold in the presence of T_4 DNA. Lysozyme activity developed in the *in vitro* system for the synthesis of protein after 30 min incubation at 37 °C. When DNA from phage T_4eG59, which has a lesion in the lysozyme gene, was used to direct lysozyme synthesis no lysozyme activity appeared in the system despite amino acid incorporation similar to that with normal T_4 DNA. The lysozyme activity thus found in the *in vitro* synthesis with T_4 DNA was completely neutralized by the antibody against purified T_4 lysozyme preparation. The antiserum inactivated the lysozyme activity observed in T_4-infected cells but not that of egg-white. Using various inhibitors of RNA and protein synthesis the dependence of lysozyme appearance *in vitro* on transcription and translation was tested. Puromycin, chloramphenicol and actinomycin D inhibited stimulation of amino acid incorporation and prevented lysozyme synthesis. DNase addition similarly inhibited the stimulation.

4.4 Penicillinase of *Bacillus licheniformis*

A hydrophobic penicillinase is reported to be synthesized *in vitro* in an extract of *B. licheniformis* 749/C [91]. An extracellular hydrophilic penicillinase protein produced by the strain of *B. licheniformis* has a molecular weight of 29000 daltons and an extremely stable secondary and tertiary structure [92]. The bacterial cells contain a membrane-bound penicillinase which is a hydrophobic protein with a molecular weight of 33000 daltons and aggregates in the absence of detergent [93, 94]. Both enzymes have been shown by genetic studies [95, 96] to be directed by a common structural gene. In fact, the exo-enzyme can be derived from the membrane-bound enzyme by limited trypsin treatment [97, 98] or incubation of the cells at pH 9.0 [99]. Membrane penicillinase differs from the exo-enzyme in that it contains a phospholipopeptide chain of amino acids at its amino terminus [100].

In vitro experiments were performed with *B. licheniformis* 749/C, a magnoconstitutive strain which produces up to 2% of its cell protein as penicillinase [95]. The cells harvested in mid-exponential phase were treated with lysozyme to weaken the cell wall in an isotonic medium and then transferred to a hypotonic medium containing Brij 58 to lyse the cell membrane. The lysate was fractionated by centrifugation.

When S-30 extract, which was obtained by centrifuging the lysate supernatant at 30000 $\times$ g for 30 min, was incubated with [^{14}C]leucine in the *in vitro* system the incorporation of the amino acid into acid-precipitable material continued linearly for 20 min and ended after 45 min. The incorporation is sensitive to chloramphenicol (50 μg/ml) and requires an exogenous energy source. The S-30 extracts usually contain a very small amount of penicillinase activity. Most of the endogenous penicillinase is removed during the course of washing and centrifugation in the extract preparation. Incubation of the *in vitro* protein synthesis system for 60 min lead to a net increase of penicillinase which corresponds to 55 units/ml. This increase can be prevented by the addition of chloramphenicol. The *in vitro* product is purified first by affinity chromatography on Sepharose 4B-cephalosporin C and then using polyacrylamide gel electrophoresis.

On a polyacrylamide gel membrane penicillinase is separated from the exopenicillinase; the latter migrates twice as fast as the former. The activity of penicillinase synthesized in the *in vitro* system migrates similarly to the membrane enzyme. However, the product obtained after incubation with chloramphenicol or with a no-energy-generating system shows much lower activity and on electrophoresis exhibits a similar distribution of enzyme to that of the unincubated sample.

When the enzyme sample synthesized *in vitro* is treated with phospholipase D [97] (to remove the phospholipid moiety, leaving serine at the amino-terminus) and then subjected to electrophoresis, the activity pattern of the sample is almost the same as that of the extracellular enzyme. The authors point out that the S-30 in the *in vitro* system of *B. licheniformis* for protein synthesis is prepared without special precautions to protect the initiation factor and accordingly, the system probably does not initiate new peptide formation but rather completes chains attached to the polysomes. They argue that due to the hydrophobicity of the phospholipid group, the penicillinase may become capable of inserting the hydrophobic group into the membrane even while the peptide chain is incomplete. This finding suggests that the hydrophobicity of the membrane enzyme plays an important role in the secretion of the penicillinase.

The nascent lipophilic enzyme may be converted to the exo-enzyme form and secreted, or pass areas of membrane not associated with secretion, therefore allowing some of the bound enzyme to be released to the exterior.

Recent findings have clarified the mode of secretion of the enzyme [101]. A hydrophobic membrane penicillinase preparation of high purity was analyzed. Comparison of the membrane enzyme with homologous enzymes indicated that the membrane enzyme has additional amino acid residues which constitute a phospholipopeptide. This was separated by treating the membrane enzyme with trypsin followed by gel filtration, leaving an enzyme which is still active but hydrophilic. Analysis of the phospholipopeptide showed that the membrane enzyme has an additional phospholipopeptide chain consisting of 25 amino acids and phosphydidyl serine at its amino terminal. Further study of the phospholipopeptide by tryptic digestion revealed that the whole sequence can be derived from a tetrapeptide (Asp(or Asn)-Glu(or Gln)-Ser-Gly) by a series of mutations, four deletions and one insertion. The mRNA corresponding to the peptide chain is deduced to have a high purine content (ca. 80%), a structure resembling poly(A). These remarkable features of this putative sequence of 75 nucleotides imply that since the poly(A) sequence at the 3′-end of mRNA from eukaryotic cells has an affinity for the membrane [102, 103] the mRNA for the penicillinase would be expected to favor translation on membrane-bound rather than free ribosomes.

4.5 α-Amylase of *Bacillus amyloliquefaciens*

The mechanism of synthesis and secretion of extracellular enzymes by microorganisms has long been of interest to microbiologists. In particular, bacterial amylase has attracted the interest of many investigators. The first *in vitro* synthesis of bacterial amylase was reported by Coleman [104]. Since the role of membrane-bound polysomes in protein synthesis *in vivo* in bacteria had been emphasized, his study focused on polysomes [105–107]. Following several surveys of the conditions for *in vitro* incorporation of amino acid, a net increase of α-amylase enzyme activity was observed in the *in vitro* system of *B. amyloliquefaciens.* Using cells harvested after 26 hr of culture, frozen-and-thawed cells, lysozyme-treated cells and membrane as well as soluble fraction of the lysed cell were tested. The abilities to incorporate [^{14}C]amino acid into protein and to increase α-amylase activity were compared on the basis of a unit amount of RNA. The net increases in the enzyme activity with cell lysate, membrane fraction and soluble fractions were 0.73, 2.62, and 0.49 unit/mg-RNA, respectively. Membrane and soluble fractions contained 11% and 89% of the total RNA, respectively. The contribution of the membrane fraction to amylase synthesis in the total lysate was estimated to be 40%, while the incorporation per unit amount of RNA was almost equal for the three samples. The role of polysomes bound to membranes in the enzyme synthesis was implied, assuming peptide chain elongation without new initiation.

As noted in the introduction, the production pattern of enzymes in microbial culture have been studied in our department [3–7, 10]. One investigation concerned the production of bacterial α-amylase [4, 6] and was extended to the *in vitro* synthesis of α-amylase [20, 108, 109].

The production pattern in a defined medium containing starch of α-amylase by *B. amyloliquefaciens* (formerly designated *B. subtilis*) strain KA63 shows a typical pattern of preferential synthesis in which the differential rate of enzyme synthesis is not correlated to the increase in cell mass (growth) [4]. When the specific rate of α-amylase synthesis (ϵ) is maximal (after around 8 h), the specific growth rate (μ), which shows its peak after 4 h, is greatly reduced (Fig. 10).

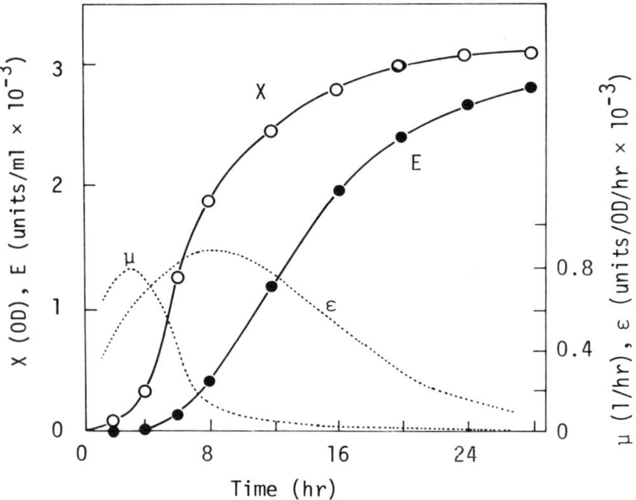

Fig. 10. Time course of α-amylase formation by *B. amyloliquefaciens*. X: cell concentration, E: enzyme rate of growth, ϵ: specific rate of enzyme formation

On examing the effect of inhibitors of protein synthesis, the increases in cell mass and α-amylase activity were found to be prevented completely by either chloramphenicol or puromycin. In contrast, the addition of actinomycin D (0.6 μg/ml), which inhibits transcription, resulted in complete inhibition of growth but allowed an increase in the amylase acitivity of 60% of the rate of the control [6] (Fig. 11). The rate of labeled amino acid incorporation into protein was found to be inhibited to a similar extent by the inhibitor, while [^{14}C]adenine incorporation was inhibited almost completely. These observations were confirmed further by additions at various culture times and suggested that the increase of enzyme-forming capacity is arrested but the expression

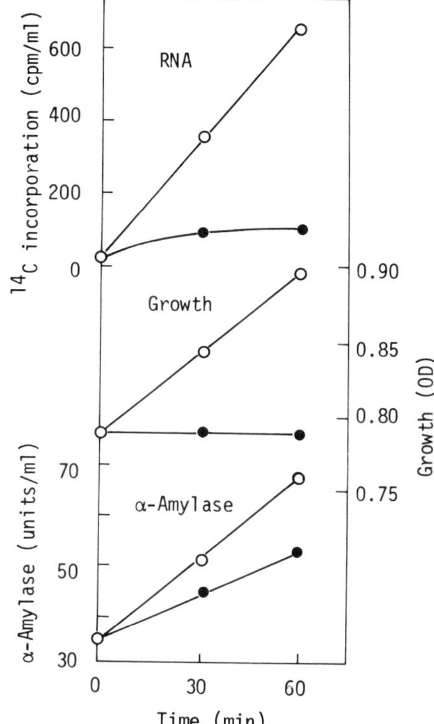

Fig. 11. Effect of actinomycin D on the production of α-amylase, cell mass and RNA. α-Amylase activity, growth and RNA were observed in the presence (•) and absence (○) of actinomycin

of pre-existing enzyme-forming capacity is not inhibited. The capacity for enzyme formation was naturally assumed to correspond to the quantity of messenger RNA for α-amylase. The extremely high stability of the messenger was suggested from the results of an actinomycin D experiment in which enzyme production continued for 140 min after the addition of the antibiotic.

The conformity of the time course of the enzyme formation to a kinetic model for the production of enzymes has been tested [4]. The model was constructed assuming that the rate of production of enzyme-forming activity (ϵ) increases proportionally with the growth rate (μ) and decreases proportionally to the first derivative of the growth rate and to the level of enzyme-forming activity in the following manner:

$$\frac{d\epsilon}{dt} = a\mu - b \cdot \frac{d\mu}{dt} - k\epsilon.$$

Conditions that were varied to test the fitness of the model with the time course of actual culture included concentration of starch, concentration of amino acids, volume of the medium and cultivation temperature. In all cases good conformity of the model to experimental data was observed and the decay constant of the enzyme-forming capacity at various temperatures agreed well with the Arrhenius equation, exhibiting an activation energy of 11.3 kcal/mol. The half-life of the capacity was estimated to be 6.3 h and 7.7 h at 35 °C and 30 °C, respectively.

The stability and accumulation of the capacity for enzyme synthesis has recently been confirmed in a series of studies on exo-enzyme formation in *B. amyloliquefaciens* by Elliot and his group [18, 110, 111]. From their results, they also concluded that formation of exo-enzymes is not inhibited by inhibitors of RNA synthesis when cells have accumulated mRNA for the enzyme in question. They propose a model for accumulation of mRNA in which the peripheral portion of the cytoplasm is the site of accumulation [18].

A number of findings indicate that a significant portion of mRNA is present in membrane-bound polysomes; these studies were reviewed recently [112–114]. Accordingly, efforts have been focused on the *in vitro* synthesis of *B. amyloliquefaciens* α-amylase based on the working hypothesis that the mRNA for the enzyme is stabilized in a membrane-bound site. As the first step in the approach to the *in vitro* synthesis, the preparation and properties of an amino acid-incorporating membrane system constructed of cell fractions of *B. amyloliquefaciens* KA63 were examined [19].

In preparing membranes capable of incorporating amino acids into protein, low temperature was employed to avoid attack by protease and nuclease, while mechanical damage was avoided by resorting to only mild agitation and suspending conditions. Cells of the strain KA63 harvested in the early log phase were lysed with egg white lysozyme at 4 °C for 60 min by suspending in a hypotonic medium. The supernatant solution (S-30) and the precipitate (crude membranes) were obtained by centrifuging the lysate at 30000 × g. S-30 was fractionated by 105000 × g centrifugation for 60 min to obtain the supernatant (S-105). Fractionated membranes were prepared by centrifuging the crude membranes on sucrose layers of 20%/60%. The membrane particles were recovered by centrifugation at 30000 × g for 20 min. About 60% of the RNA was recovered in the supernatant. The amount of membrane-bound RNA and DNA recovered in the crude membranes was 12–16% and 14–17% of cell contents, respectively.

Figure 12 shows the distribution of the DNA, RNA and the protein of the crude membranes between the layers of sucrose after centrifugation. Sharp peaks of DNA, RNA, protein and [^{14}C]glycerol, which was added as a marker for lipid, are centered in a dense layer around the interface.

A cell-free system incorporating amino acid was constructed making a slight modification of the method of Matthaei and Nirenberg [1, 2]. The cell-free system contained a Tris-HCl buffer, Mg-acetate, NH_4Cl, β-mercaptoethanol, ATP, GTP, creatine phosphate, creatine phosphokinase, 20 amino acids, a [^{14}C]amino acid mixture, S-30 or S-105 and purified membranes. The characteristics of the amino acid-incorporation system are summarized in Table 2. The [^{14}C]amino acid incorporation into hot TCA-insoluble protein in the *in vitro* system is insentive to KCN and not inhibited by actinomycin D and rifampicin, indicating that the reaction is directed by mRNA originally present in the membrane preparation. The time courses of protein synthesis in the system are shown in Fig. 13 in which the activities of crude and purified membranes are compared. With a combination of membranes and supernatant fraction, the incorporation of amino acid increased. Purified membranes and S-105 do not show protein synthesis individually but only when both are jointly present. Since S-105 is thought to be free from mRNA

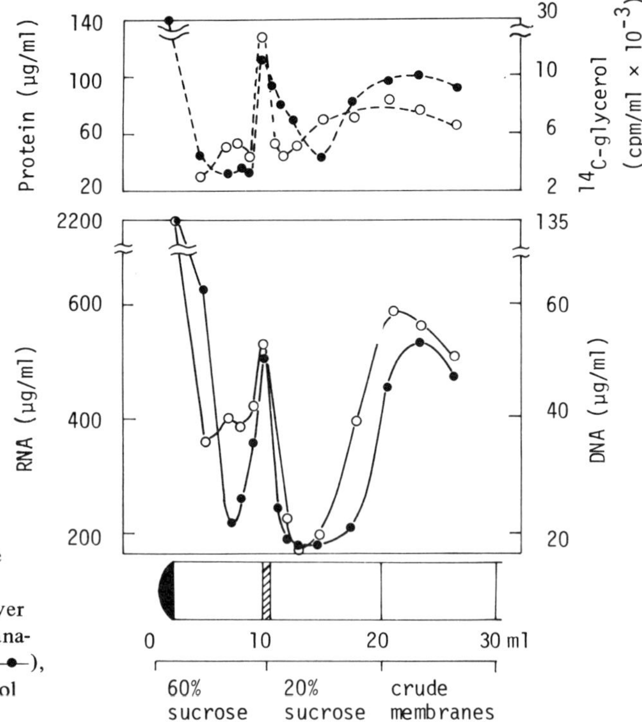

Fig. 12. Fractionation of crude membranes by centrifugation through 20%/60% sucrose bilayer media. One-ml fractions were analyzed for RNA (–○–), DNA (–●–), protein (---○---) and [^{14}C]glycerol (---●---)

Table 2. Characteristics of radioactive amino acid incorporation by the cell-free system of *B. amyloliquefaciens*

Addition or omission	Incorporation (%)	
	S-30 system	S-105 plus membrane system
None	100	100
– Membrane	–	0
– S-105 or S-30	0	0
– ATP, GTP, PC[a] and PC kinase	0	0
+ KCN (100 µg/ml)	86	108
+ Chloramphenicol (100 µg/ml)	13	15
+ Puromycin (50 µg/ml)	14	–
+ Rifampicin (10 µg/ml)	85	–
+ Actinomycin D (10 µg/ml)	77	97

[a] Phosphocreatine.

Reaction mixtures were incubated for 60 min at 30 °C.

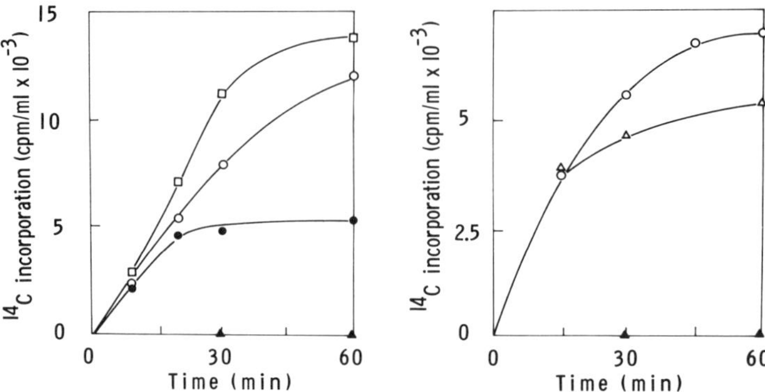

Fig. 13. Activities of crude and fractionated membranes in incorporating amino acid into protein. In the figure to the left, the symbols are defined as follows: □: crude membranes with S-30, ○: purified membranes with S-30, ●: S-30 without membranes, ▲: crude membranes without supernatant fraction. In the figure to the right, the symbols are defined as: ○: purified membranes with S-30, △: purified membranes with S-105, ▲: S-105 without membranes

and ribosomes, the membranes would be the source of polysomes, that are most likely bound to the membranes. In fact, the polysomes were not removed by repeated washing with buffer solution. On the other hand, S-30 showed protein synthesis even without addition of membranes, suggesting the presence of free polysomes in S-30.

The factors limiting the rate and extent of protein synthesis in the membrane system were surveyed. With an increase in the amount of S-105 the initial rate of incorporation also increases but the maximum amount incorporated is not affected. The maximum amount incorporated was proportional to the amount of membranes although the initial rate was only slightly affected. To investigate the factors limiting the continuation of the incorporation, fresh membranes were added to the reaction system when the protein synthesis had reached a plateau (Fig.14). Initially, synthesis resumed at almost the same rate as that of the fresh reaction mixture and the amount of amino acid further incorporated corresponded to the amount of membranes added. This observation suggests that factors present in S-105 are stable even after the cessation of incorporation and that the reaction ceased due to a loss of membrane activity. Two possibilities may be considered for this loss of activity: one concerns the release of ribosomes from the membranes, the other the decay of mRNA.

To clarify these possibilities, purified ribosomes were added after cessation of protein synthesis. Ribosomes prepared by washing with NH_4Cl showed three peaks corresponding to 70 S, 50 S, and 30 S, when using a sucrose gradient analysis. When the protein-synthesizing activity of the washed ribosomes was tested without membranes, the incorporation of [^{14}C]phenylalanine was observed in the presence of polyuridylic acid but not in the absence of polyuridylic acid. The washed ribosomes are active and mRNA is not contained either in the ribosome preparation or S-105.

When the reaction had reached a plateau in the complete reaction mixture, the addition of 100 A_{260} units/ml of ribosomes, at 30 min, caused resumed incorporation. From

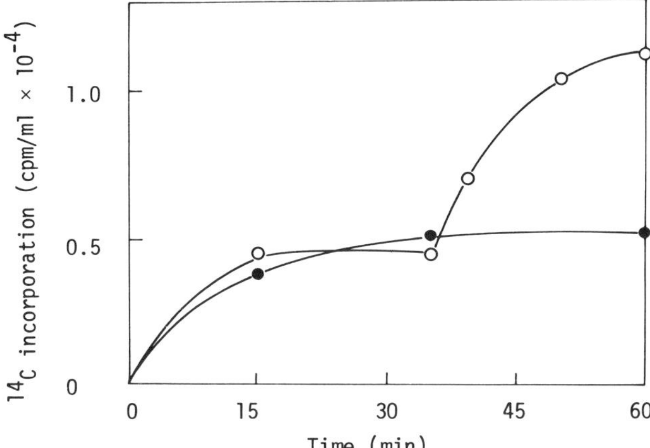

Fig. 14. Protein synthesis resumed as the result of the addition of purified membranes. After 35 min of reaction about twice the initial amount of membranes was added

these observations it may be concluded that the mRNA activity remained even after cessation of protein synthesis and that ribosomes released from membrane-bound polysomes in the course of the reaction cannot account for new peptide synthesis because of their dilution in the reaction mixture. Therefore, the addition of an excess amount of ribosomes compensate for the lack of ribosomes. The effect of subunit addition was also examined. The addition of 30 S subunits stimulated the incorporation while 50 S subunits repressed the incorporation. The stimulation by 30 S subunits suggests the possibility that polysomes are attached to membranes through larger subunits, as in the model proposed for mammalian cells, and that smaller subunits attach more loosely. To elucidate whether the membranes are the site of protein synthesis a sucrose gradient sedimentation analysis was performed. The reaction mixtures during the course of reaction were analyzed through a 15–30% sucrose gradient. After 30 min of reaction an increase in radioactivity was observed among the fractions containing membranes. In the fractions containing free polysomes no significant activity or UV absorption was detected. The addition of chloramphenicol or puromycin prevented the increase of radioactivity. These results suggest that translation occurs in the membranes and that membrane-bound polysomes dissociate to free ribosomes and their subunits during the reaction. The above results show that the protein synthesis in the cell-free system from *B. amyloliquefaciens* KA63 is directed exclusively by endogenous mRNA bound to the membranes.

As a further step, an examination was made to determine whether or not a particular secretory protein such as α-amylase is synthesized preferentially on membrane-bound poylsomes [108]. Immunological determination was applied using a mutant strain (AL222) which produces less α-amylase of the same structure. The immunochemical homology was confirmed by a double diffusion test using anti-amylase serum prepared against liquefying α-amylase of *B. subtilis* in a rabbit. The anti-amylase reacted with strain KA63 amylase in the same way as the mutant amylase. A neutralization test also proved the immunological homology of parental and mutant α-amylases. Based on these facts, the amount of α-amylase in the products of the cell-free system was assayed immunologically. After incubation, the reaction mixtures in the cell-free system were centrifuged and the supernatant was used in the immunological assay. The immunoprecipitate formed was washed with saline solution and its radioactivity determined. In the system containing membranes from a 3 h culture of strain KA63, a considerable level of radioactivity was observed in the immunoprecipitate, corresponding to 16.2–17.2% of the total radioactivity incorporated into the hot TCA-insoluble material (Table 3).

Table 3. Incorporation of [^{14}C]amino acid into anti-amylase precipitate by *in vitro* synthesis of *B. amyloliquefaciens*

Strain	Exp.	System	Incorp. into total protein (A)	Incorp. into anti-amylase precipitate (B)	Ratio (B/A)
			(cpm/ml)	(cpm/ml)	(%)
KA63	1	Membranes	15,080	2,600	17.2
		S-30 supnt.	49,810	360	0.7
	2	Membranes	9,890	1,600	16.2
		Membranes (6 h)	4,050	2,560	63.2
		S-30 supnt.	35,780	600	1.7
AL222	3	Membranes	18,100	ND	–
		S-30 supnt.	59,360	ND	–

Further, when the membranes obtained from cells producing α-amylase at the maximal rate in a 6 h culture were employed, the recovery of radioactivity in the immunoprecipitate was 63%. In contrast, in the S-30 system, which is considered to contain free polysomes, the radioactivity in the immunoprecipitate was very low. Radioactivity was not observed in the immunoprecipitate of both systems from strain AL222, despite the fact that these cell-free system had sufficient activity to incorporate amino acid. Considering the fact that strain AL222 has reduced productivity of α-amylase, these results suggest that radioactivity in the immunoprecipitate represents the amount of α-amylase in the products of the cell-free system.

In summary, results obtained in the study suggest that α-amylase is synthesized on membrane-bound polysomes and that mRNA specific for α-amylase is accumulated in the membrane region, escaping from degradation in the course of culture.

5. Concluding Remarks

In this article *in vitro* synthesis of enzymes was discussed with special attention to microbial enzymes and mRNA molecules. The importance of the role of mRNA in the synthesis of enzymes has been pointed out. It was shown that the lysozyme of T_4 phage is synthesized *in vitro* under the direction of an isolated RNA preparation and that the mRNA for the synthesis of the lipoprotein in the outer membrane of *E. coli* was purified 250-fold. The analysis of the amino acid sequence at the amino terminus of hydrophobic exopenicillinase bound to the cell wall of *E. coli* indicates that the mRNA for the penicillinase is enriched with purine bases (80%) at the 5′-terminal sequence and the purine-rich sequence might contribute in binding the mRNA to membranes as the poly(A) sequence has been assumed to bind eukaryotic mRNA. Evidence was shown that *in vitro* synthesis of α-amylase of *B. amyloliquefaciens* takes place on ribosomes on the mRNA for the enzyme, which are specifically enriched in the membrane fraction. These facts are compatible with the hypothesis that specific proteins or enzymes secreted into the medium by bacterial cells are synthesized on polysomes containing specific mRNAs bound to the membranes of living cells. With respect to the affinity of polysomes for the membrane, two mechanisms have been proposed. One is that lipid groups covalently linked to the amino terminus of the nascent peptide chain of the secretory protein on polysomes work as an anchor for membranes; the other is that a specific sequence of mRNA close to the 3′-terminus is bound to a specific protein species located in the membrane by an affinity similar to that between a repressor and DNA at an operator site. Experimental evidence supporting both possibilities has been presented in different systems. The high stability of mRNA for the secretory enzyme may be attributed to protection from ribonuclease action caused by the binding of polysomes to the membrane.

The production processes of microbial enzymes can be analyzed and improved by estimating the parameters related to the formation and the decay of enzyme-synthesizing systems (mRNA) and by obtaining information on the extent and types of regulatory controls such as induction (or derepression) and catabolite repression (or anabolite repression).

The finding that mRNA for a specific enzyme is accumulated in membrane fractions indicates the possibility of preparation of the mRNA. Recent progress in techniques has made purification of biologically active preparations of mRNA possible by applying immunological precipitation of polysomes containing nascent peptide chains and chromatography. Techniques of reverse transcription have been shown effective in preparing DNA, complementary to isolated mRNA in several cases [115–117]. Information concerning the recombination of heterogeneous DNAs and the cloning of DNA is rapidly increasing [118–122]. Efforts along these lines will lead to the preparation and the cloning of gene DNA for a specific protein or enzyme. Studies at the level of chimeric recombinant DNA are expected in the near future to provide information required for preferential synthesis *in vivo* of certain important enzymes.

References

1. Matthaei, J. H., Nirenberg, M. W.: Proc. Nat. Acad. Sci. USA **47**, 1580 (1961).
2. Nirenberg, M. W., Matthaei, J. H.: Proc. Nat. Acad. Sci. USA **47**, 1588 (1961).
3. Terui, G., Okazaki, M., Kinoshita, S.: J. Ferment. Technol. **45**, 497 (1967).
4. Kinoshita, S., Okada, H., Terui, G.: J. Ferment. Technol. **45**, 504 (1967).
5. Okazaki, M., Terui, G.: J. Ferment. Technol. **45**, 1147 (1967).
6. Kinoshita, S., Okada, H., Terui, G.: J. Ferment. Technol. **46**, 427 (1968).
7. Shinmyo, A., Okazaki, M., Terui, G.: J. Ferment. Technol. **46**, 733 (1968).
8. Terui, G.: Dynamics of microbial cultures. In: Outline of lectures for International post-graduate university course in microbiology in Japan. 1973 p. 160.
9. Tahara, T., Shinmyo, A., Enatsu, T., Terui, G.: J. Ferment. Technol. **52**, 517 (1974).
10. Shinmyo, A., Okazaki, M., Terui, G.: J. Ferment. Technol. **46**, 1000 (1968).
11. Garcia-Martinez, D. V., Ogawa, T., Shinmyo, A., Enatsu, T.: J. Ferment. Technol. **52**, 378 (1974).
12. Doi, S., Shinmyo, A., Enatsu, T., Terui, G.: J. Ferment. Technol. **51**, 768 (1973).
13. Jacob, F., Monod, J.: J. Mol. Biol. **3**, 318 (1961).
14. Kepes, A.: Biochim. Biophys. Acta **138**, 107 (1967).
15. Kennell, D., Bicknell, I.: J. Mol. Biol. **74**, 21 (1972).
16. Cybis, J., Weglenski, P.: Eur. J. Biochem. **30**, 262 (1972).
17. Enatsu, T., Umezaki, Y., Kise, S., Muth, T.: In: Abstracts of papers, Fifth International Fermentation Symposium. Berlin, Bonn, Westkreuz-Druckerei und Verlag, 1976 p. 164.
18. Both, G. W., McInnes, J. L., Hanlon, J. E., May, B. K., Elliott, W. H.: J. Mol. Biol. **67**, 199 (1972).
19. Nimomiya, Y., Imanishi, T., Shinmyo, A., Enatsu, T., Terui, G.: J. Ferment. Technol. **53**, 189 (1975).
20. Shinmyo, A., Terui, G.: J. Ferment. Technol. **48**, 519 (1970).
21. Tahara, T., Kotani, H., Shinmyo, A., Enatsu, T.: J. Ferment. Technol. **53**, 409 (1975).
22. Hendler, R. W.: Protein biosynthesis and membrane biochemistry. John Wiley & Sons Inc., New York, London 1968 pp. 344.
23. Lengyel, P., Söll, D.: Bacteriol. Rev. **33**, 264 (1969).
24. Lucas-Lenard, J., Lipmann, F.: Ann. Rev. Biochem. **40**, 409 (1971).
25. Bosch, L. (Ed.): The mechanism of protein synthesis and its regulation. (Frontiers of Biology Vol. 27) North-Holland Publish. Co., Amsterdam, London, 1972 pp. 590.
26. Arnstein, H. R. V. (Ed.): Synthesis of amino acids and proteins. (MTP international review of science, Biochemistry series 1, Vol. 7) Butterworths, London, University Park Press, Baltimore, 1975 pp. 416.
27. Haselkorn, R., Rothman-Denes, L. B.: Ann. Rev. Biochem. **42**, 397 (1973).
28. Geideschek, E. P., Haselkorn, R.: Ann. Rev. Biochem. **38**, 647 (1969).
29. Lodish, H. F.: Ann. Rev. Biochem. **45**, 39 (1976).
30. Lewin, B.: Synthesis of protein. In: Gene expression Vol. 1, John Wiley & Sons, London, 1974 p. 3.
31. Hirashima, A., Wang, S., Inouye, M.: Proc. Nat. Acad. Sci. USA **71**, 4149 (1974).
32. Braun, V., Bosch, V.: Proc. Nat. Acad. Sci. USA **69**, 970 (1972).
33. Braun, V. Bosch, V.: Eur. J. Biochem. **28**, 51 (1972).
34. Hantke, K., Braun, V.: Eur. J. Biochem. **34**, 284 (1973).
35. Braun, V., Siegling, J.: Eur. J. Biochem. **13**, 336 (1970).
36. Braun, V., Wolff, H.: Eur. J. Biochem. **14**, 387 (1970).
37. Inouye, M., Shaw, J., Shen, C.: J. Biol. Chem. **247**, 8154 (1972).
38. Hirashima, A., Wu, H. C., Vankateswaran, P. C., Inouye, M.: J. Biol. Chem. **248**, 5654 (1973).
39. Hirashima, A., Childs, G., Inouye, M.: J. Mol. Biol. **79**, 373 (1973).
40. Takeishi, K., Yasumura, M., Pirtle, R., Inouye, M.: J. Biol. Chem. **251**, 6259 (1976).
41. Uenoyama, K., Ono, T.: J. Mol. Biol. **65**, 75 (1972).
42. Brawerman, G., Mendecki, J., Lee, S. Y.: Biochemistry **11**, 637 (1972).

43. Aviv, H., Leder, P.: Proc. Nat. Acad. Sci. USA **69**, 1408 (1971).
44. Oda, K., Joklik, W. K.: J. Mol. Biol. **27**, 395 (1967).
45. Penman, S.: J. Mol. Biol. **17**, 117 (1966).
46. Brawerman, G.: Ann. Rev. Biochem. **43**, 621 (1974).
47. Perry, R. P., La Torre, J., Kelley, D. E., Greenberg, J. R.: Biochim. Biophys. Acta **262**, 220 (1972).
48. Crystal, R. G., Nienhuis, A. W., Elson, N. A., Anderson, W. F.: J. Biol. Chem. **247**, 5357 (1972).
49. Adesnik, M., Salditt, M., Thomas, W., Darnell, J. E.: J. Mol. Biol. **71**, 21 (1972).
50. Schechter, I.: Proc. Nat. Acad. Sci. USA **70**, 2256 (1973).
51. Krystosek, A., Cawthon, M. L., Kabat, D.: J. Biol. Chem. **250**, 6077 (1975).
52. Kabat, D.: J. Biol. Chem. **250**, 6085 (1975).
53. Lockard, R. E., Lingrel, J. B.: J. Biol. Chem. **247**, 4174 (1972).
54. Palmiter, R. D.: J. Biol. Chem. **248**, 2095 (1973).
55. Adamson, S. D., Herbert, D., Godchaux, W.: Arch. Biochem. Biophys. **125**, 671 (1968).
56. Zucker, W. V., Shulman, H. M.: Proc. Nat. Acad. Sci. USA **59**, 582 (1968).
57. Maxwell, C. R., Kamper, C. S., Rabinovitz, M.: J. Mol. Biol. **58**, 317 (1971).
58. Adamson, S. D., Yau, P. M.-P., Herbert, E., Zucker, W. V.: J. Mol. Biol. **63**, 247 (1972).
59. Gross, M., Rabinovitz, M.: Proc. Nat. Acad. Sci. USA **69**, 1565 (1972).
60. Mizumi, S., Fisher, J. M., Rabinovitz, M.: Biochim. Biophys. Acta **272**, 683 (1972).
61. Fan, H., Penman, S.: J. Mol. Biol. **60**, 655 (1970).
62. Lodish, H. F., Jacobson, M.: J. Biol. Chem. **247**, 3622 (1972).
63. Bruns, G. P., London, I. M.: Biochem. Biophys. Res. Comm. **18**, 236 (1965).
64. Rabinovitz, M., Freedman, M. L., Fisher, J. M., Maxwell, C. R.: Cold Spring Harbor Symp. Quant. Biol. **34**, 567 (1969).
65. Howard, G. A., Adamson, S. D., Herbert, E.: Biochim. Biophys. Acta **213**, 237 (1970).
66. Hunt, T., Vanderhoff, G., London, I. M.: J. Mol. Biol. **66**, 471 (1972).
67. Grayzel, A. I., Horchner, P., London, I. M.: Proc. Nat. Acad. Sci. USA **55**, 650 (1966).
68. Waxman, H. S., Rabinovitz, M.: Biochim. Biophys. Acta **129**, 369 (1966).
69. Legon, S., Jakson, R. J., Hunt, T.: Nature New Biology **241**, 150 (1973).
70. Ranu, R. S., Levin, D. H., Delaunay, J., Ernst, V., London, I. M.: Proc. Nat. Acad. Sci. USA **73**, 2720 (1976).
71. Levin, D. H., Ranu, R. S., Ernst, V., Fifer, M. A., London, I. M.: Proc. Nat. Acad. Sci., USA **72**, 4849 (1975).
72. Levin, D. H., Ranu, R. S., Ernst, V., London, I. M.: Proc. Nat. Acad. Sci. USA **73**, 3112 (1976).
73. Kramer, G., Cimadevilla, J. M., Hardesty, B.: Proc. Nat. Acad. Sci. USA **73**, 3078 (1976).
74. Redman, C. M., Siekevitz, P., Palade, G. E.: J. Biol. Chem. **241**, 1150 (1966).
75. Palade, G. E.: Microsomal particles and protein synthesis. In: First Symposium of the Biophysical Society. Roberts, R. B. (Ed.), Pergamon Press Inc., New York, 1958 p. 36.
76. Caro, L. G., Palade, G. E.: J. Cell. Biol. **20**, 473 (1964).
77. Siekevitz, P., Palade, G. E.: J. Biophys. Biochem. Cytol. **7**, 619 (1960).
78. Tanaka, T., Ogata, K.: J. Biochem. **70**, 693 (1971).
79. Khairul Bashar, S. A. M. K., Parish, J. H., Brown, M.: Biochem. J. **123**, 355 (1971).
80. Salser, W., Gesteland, R. F., Bolle, A.: Nature **215**, 588 (1967).
81. DeVries, J. K., Zubay, G.: Proc. Nat. Acad. Sci. USA **57**, 1010 (1967).
82. Imai, M., Yura, T., Marushige, K.: Biochem. Biophys. Res. Comm. **11**, 270 (1963).
83. Tsugita, A., Inouye, M.: J. Mol. Biol. **37**, 201 (1968).
84. Inouye, M., Tsugita, A.: J. Mol. Biol. **22**, 193 (1966).
85. Salser, W., Gestland, R. F., Ricard, B.: Cold Spring Harbor Symp. Quant. Biol. **34**, 771 (1969).
86. Wilhelm, J. M., Haselkorn, R.: Proc. Nat. Acad. Sci. USA **65**, 388 (1970).
87. Lacroute, F., Stent, G.: J. Mol. Biol. **35**, 165 (1968).
88. Kepes, A.: Biochim. Biophys. Acta **76**, 293 (1968).
89. Gold, L. M., Schweiger, M.: Proc. Nat. Acad. Sci. USA **62**, 892 (1968).
90. Schweiger, M., Gold, L. M.: Proc. Nat. Acad. Sci. USA **63**, 1351 (1969).
91. Dancer, B. N. Lampen, J. O.: Biochem. Biophys. Res. Comm. **66**, 1357 (1975).

in *Bacillus licheniformis* and their significance for the secretion process. In: Biomembranes, Mason, L. A. (Ed.), Plenum Publ. Corp., New York, 1971 Vol. 2, p. 211.
93. Crane, L. J., Lampen, J. O.: Arch. Biochem. Biophys. **160**, 655 (1974).
94. Sargent, M. G., Lampen, J. O.: Arch. Biochem. Biophys. **136**, 167 (1970).
95. Dubnau, D., Pollock, M. R.: J. Gen. Microbiol. **41**, 7 (1965).
96. Sherratt, D. J., Collins, J. F.: J. Gen. Microbiol. **76**, 217 (1973).
97. Sawai, T., Crane, L. J., Lampen, J. O.: Biochem. Biophys. Res. Comm. **53**, 523 (1973).
98. Sawai, T., Lampen, J. O.: J. Biol. Chem. **249**, 6288 (1974).
99. Sargent, M. G., Lampen, J. O.: Proc. Nat. Acad. Sci. USA **65**, 962 (1970).
100. Yamamoto, S., Lampen, J. O.: J. Biol. Chem. **250**, 3212 (1975).
101. Yamamoto, S., Lampen, J. O.: Proc. Nat. Acad. Sci. USA **73**. 1457 (1976).
102. Milcarek, C., Penman, S.: J. Mol. Biol. **89**, 327 (1974).
103. Lande, M. A., Adesnik, M., Sumida, M., Tashiro, Y., Sabatini, D. D.: J. Cell Biol. **65**, 513 (1975).
104. Coleman, G.: Biochem. J. **116**, 753 (1970).
105. Coleman, G.: Biochim. Biophys. Acta **174**, 395 (1969).
106. Coleman, G.: Biochim. Biophys. Acta **182/80** (1969).
107. Coleman, G.: Biochem. J. **112**, 544 (1969).
108. Ninomiya, Y., Imanishi, T., Shinmyo, A., Enatsu, T.: J. Ferment. Technol. **54**, 374 (1976).
109. Ninomiya, Y., Imanishi, T., Shinmyo, A., Enatsu, T.: J. Ferment. Technol. **54**, 459 (1976).
110. Gould, A. R., May, B. K., Elliott, W. H.: J. Mol. Biol. **73**, 213 (1973).
111. Glenn, A. R., Both, G. W., McInnes, J. L., May, B. K., Elliott, W. H.: J. Mol. Biol. **73**, 221 (1973).
112. Glenn, A. R.: Ann. Rev. Microbiol. **30**, 41 (1976).
113. Sabatini, D. D., Kreibich, G.: Functional specialization of membrane-bound ribosomes in eukaryotic cells, In: The enzymes of biological membranes, Vol. 2, Plenum Press, New York, London 1976 P. 535.
114. Lingrel, J. B.: The translation of messenger RNA in cell-free systems, In: Synthesis of amino acids and proteins, Arnstein, H. R. V. (Ed.) Butterworth Co., London & University Park Press, Baltimore, 1975 p. 295.
115. Rougeon, F., Mach, B.: Proc. Nat. Acad. Sci. USA **73**, 3418 (1976).
116. Anderson, J. N., Schimke, R. T.: Cell **7**, 331 (1976).
117. Efstratiadis, A., Maniatis, T., Kafatos, F. C., Jeffrey, A., Vournaiks, J. N.: Cell **4**, 367 (1975).
118. Kramer, R. A., Cameron, J. R., Davis, R. W.: Cell **8**, 227 (1976).
119. Maniatis, T., Kee, S. G., Kafatos, F. C.: Cell **8**, 163 (1976).
120. Collins, C. J., Jackson, D. A., deVries, F. A. J.: Proc. Nat. Acad. Sci. USA **73**, 3838 (1976).
121. Hershfield, V., Boyer, H. W., Yanofsky, C., Lovett, M. A., Helinski, D. R.: Proc. Nat. Acad. Sci. USA **71**, 3455 (1974).
122. Polisky, B., Bishop, R. J., Gelfand, D. H.: Proc. Nat. Acad. Sci. USA **73**, 3900 (1976).

Molecular Biology, Biochemistry and Biophysics

Editors: A. Kleinzeller, G. F. Springer, H. G. Wittmann

Volume 18: H. Kersten, W. Kersten
Inhibitors of Nucleic Acid Synthesis
Biophysical and Biochemical Aspects
1974. 73 figures. IX, 184 pages
ISBN 3-540-06825-2

Volume 19: M. B. Mathews
Connective Tissue
Macromolecular Structure and Evolution
1975. 31 figures. XII, 318 pages
ISBN 3-540-07068-0

Volume 20: M. A. Lauffer
Entropy-Driven Processes in Biology
Polymerization of Tobacco Mosaic Virus Protein and Similar Reactions
1975. 90 figures. X, 264 pages
ISBN 3-540-06933-X

Volume 21: R. C. Burns, R. W. F. Hardy
Nitrogen Fixation in Bacteria and Higher Plants
1975. 27 figures. X, 189 pages
ISBN 3-540-07192-X

Volume 22: H. J. Fromm
Initial Rate Enzyme Kinetics
1975. 88 figures, 19 tables, X, 321 pages
ISBN 3-540-07375-2

Volume 23: M. Luckner, L. Nover, H. Böhm
Secondary Metabolism and Cell Differentiation
1977. 52 figures, 7 tables. VI, 130 pages
ISBN 3-540-08081-3

Volume 24
Chemical Relaxation in Molecular Biology
Editors: I. Pecht, R. Rigler
1977. 141 figures, 50 tables.
XVI, 418 pages
ISBN 3-540-08173-9

Volume 25
Advanced Methods in Protein Sequence Determination
Editor: S. B. Needleman
1977. 97 figures, 25 tables.
XII, 189 pages
ISBN 3-540-08368-5

Volume 26: A. S. Brill
Transition Metals in Biochemistry
1977. 49 figures, 18 tables.
VIII, 186 pages
ISBN 3-540-08291-3

Volume 27
Effects of Ionizing Radiation on DNA
Physical, Chemical and Biological Aspects
Editors: A. J. Bertinchamps (Coordinating Editor), J. Hüttermann, W. Köhnlein, R. Téoule
1978. 74 figures, 47 tables.
XXII, 386 pages
ISBN 3-540-08542-4

Volume 28: A. Levitzki
Quantitative Aspects of Allosteric Mechanisms
1978. 13 figures, 2 tables.
Approx. 125 pages
ISBN 3-540-08696-X